THE FACTS ON FILE MARINE SCIENCE HANDBOOK

SCOTT McCUTCHEON
and
BOBBI McCUTCHEON

Facts On File, Inc.

Facts On File, Inc.
132 West 31st Street
New York NY 10001

Library of Congress Cataloging-in-Publication Data

McCutcheon, Scott.
The Facts on File marine science handbook / Scott McCutcheon and Bobbi McCutcheon.
p. cm. — (the Facts on File science handbooks)
Includes bibliographical references and index.
ISBN 0-8160-4812-6 (acid-free paper)
1. Marine sciences — Handbooks, manuals, etc. I. McCutcheon, Bobbi. II. Title, III. Series.
GC24.M43 2003
551.46 — dc21 2003000275

Facts On File books are available at special discounts when purchased in bulk quantities for businesses, associations, institutions, or sales promotions. Please call our Special Sales Department in New York at 212/967-8800 or 800/322-8755.

You can find Facts On File on the World Wide Web at http://www.factsonfile.com

Cover design by Cathy Rincon
Illustrations by Bobbi McCutcheon and Scott McCutcheon, © Facts On File

Printed in the United States of America

VB Hermitage 10 9 8 7 6 5 4 3 2 1

This book is printed on acid-free paper.

CONTENTS

ACKNOWLEDGMENTS

We the authors would like to thank the National Oceanic and Atmospheric Administration (NOAA) and their staff members for generously providing much of the information and many of the photos herein.

Also, we extend our gratitude to the Alaska Lighthouse Association for freely donating their time and materials. Additionally, we would like to express our deepest appreciation Frank K. Darmstadt, Senior Editor, Facts On File, for his invaluable guidance throughout the entire book-making process, Copy Editor Susan Thornton, and the entire Facts On File staff for their time and energies in the development of this project.

PREFACE

Who are we? Where do we come from? These are two of humankind's most asked questions. Though scientists differ on the answer, one that still yearns to be solved irrefutably, there is one fact we know with certainty: where there is water there is life. Marine science is the ultimate requisite when exploring and understanding the world in which we live. Within our solar system, Earth is the only body to have not only enough water to support life, but also enough to foster virtually millions of species, many of which are still waiting to be revealed.

One finds it fascinating that we as human beings live each day above the unassuming seas while beneath them in reality lies a hidden ecosystem teeming with creatures and vegetation so bizarre they really seem to be inhabitants of another world. Some consider space the final frontier; yet the oceans of Earth are just as mysterious, and as far as we have come in understanding them, there is still so much to be discovered.

What would we give to know intimately and thoroughly our planet's aquatic habitats? Descending to extreme depths in the vast waters of our world's front yard demands equipment and technology as complex and advanced as those that travel into outer space. And what *is* our primary goal of nearly each rocket mission that powers its way into our solar system to rendezvous with another world? Primarily to search for one important element: water. And if on another world, one as close as Mars or the moons of Jupiter, we were to find liquid water, and with it perhaps ancient or newly forming life, it would undoubtedly generate a staffed mission. What if the most important person on that future mission were the marine biologist? What if that biologist were you? But how can we begin to understand water-generated life on other worlds without first grasping the secrets of our own?

This primer on marine science is the perfect start. Within these pages designed especially for you can be found a selected list of easy-to-follow terms, descriptions, drawings, biographies of the great people who have made historic contributions in marine science, a chronology of exploration and research, and straightforward charts to augment the terms and definitions.

There appear to be many very separate areas of science, but as we grow and each distinct caste becomes clear, we find that all science is tied together. Given the fact that three-quarters of Earth's surface is covered in water, one cannot wonder why marine science is at the forefront of curiosity. We believe that in order to conquer the mysteries of our world, humankind must look first toward the sea.

INTRODUCTION

GLOSSARY

This list contains almost 1,500 entries: alphabetically listed terms and their definitions used in the field of marine science, often accompanied by a graph or chart to clarify meanings. A well-equipped and easily understood glossary within arm's reach is essential for any field of study.

BIOGRAPHIES

Throughout the history of the systematic study of the oceans, many great people have stood on the shoulders of their eminent predecessors to broaden knowledge about our marine environment. These early to modern sailors and scientists began the process that allows today's mind to understand more fully our greatest nurturer of life, the oceans. Listed here are close to 250 of the most renowned of those people, along with many less familiar individuals who have earned recognition for their selfless efforts and principal contributions in the study of marine science.

CHRONOLOGY

On an evolutionary scale, the discovery of science and technology has evolved on Earth in the twinkle of an eye, but in human terms, for the most part, it has not been what anyone could call fast paced. Patience must sometimes be employed; yet the rewards are great, and each exciting breakthrough inevitably causes change, however gradual. Nearly 200 entries are listed here, providing a general historical account of oceanic explorations and marine research spanning 6,000 years, events that have helped revolutionize civilized humanity's understanding of the sea.

CHARTS AND TABLES

Reliable information is the key to basic knowledge, and few methods work better than use of charts or graphs in enhancing understanding of new material. For quick reference this primer offers a valuable resource of charts and tables dealing with the important, fundamental facts of our aquatic environment such as the tides in relationship to the Earth, Moon, and Sun; major shipping routes; examples of ocean food chains; and global wind circulation and oceanic surface currents.

SECTION ONE

GLOSSARY

abaft Behind. A nautical term meaning toward the stern (rear) of a boat.

abalone A primitive, univalve gastropod found on the rocky shores of warm and temperate seas, except the western Atlantic Ocean. We value them for their large, edible foot and the mother-of-pearl lining in their shells, which commonly is used to make decorative items such as jewelry. Abalones constitute the family Haliotidae, in the order Archaeogastropoda, class Gastropoda, phylum Mollusca, kingdom Animalia.

abdomen In a TETRAPOD, the portion of its body between the thorax and pelvic girdle. In an ARTHROPOD, the region of its body behind the thorax. It contains the visceral (internal) organs.

abeam A nautical term meaning at a right angle to a ship's keel, yet not on the ship.

abiotic The nonexistence of life, or nonbiological.

abyssal hills Low, undersea hills characteristic only of the deep sea. Mainly, they rise from about 10 to 750 feet (3–228 m) high and range anywhere from a few hundred feet to a few miles wide.

abyssal plain

abyssal plain Also known as basins or ocean basins, these are flat expanses of the deep seafloor that slope less than one part in 1,000 and typically exist between the base of abyssal hills and the continental slope. Many basins and plains are identified in this book.

abyssopelagic zone The lightless area of open ocean beneath the BATHYPELAGIC ZONE, existing at about 12,000 to 20,000 feet (3,657–6,096 m). Examples of sea creatures adapted to such depths are deep-water eels, such as gulpers; deep-sea anglers; and stomiatoids.

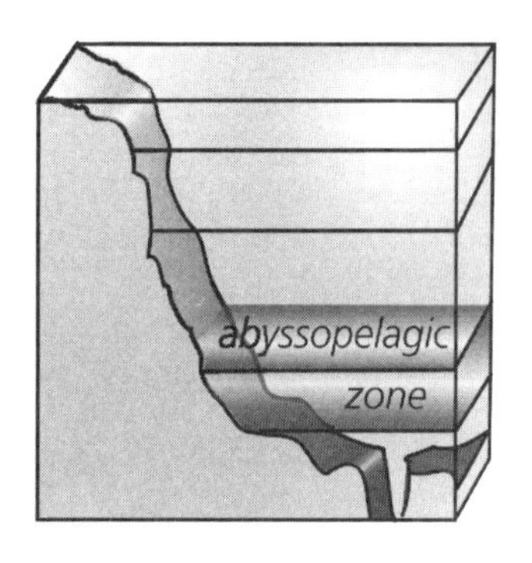

abyssopelagic zone

acanthodians A group of the earliest known true jawed fishes from the Lower Silurian to the Lower Permian period. They were sharklike and had spines on their bodies.

acetosporan A simple PROTOZOAN, parasitic on marine invertebrates, that is characterized as having multicellular spores. It can cause diseases and mass mortality.

acicula Needlelike supporting bristle existing in the PARAPODIA of some POLYCHAETES (worms). It can also exist in the form of crystals of certain minerals.

acidic A term descriptive of a substance having acidlike qualities and possessing a relatively large number of hydrogen ions; having a pH less than 7.0.

acoustic oceanography A study of ocean topology by way of sound waves, ranging from the earliest use of depth sounding to today's use of sonar.

acoustics The technology and science of sound, its generation, transmission, reception, and effect. In oceanography, acoustics plays an important role because sound waves are capable of traveling long distances unimpeded through water, bouncing off an object, and then returning to the source.

acoustic tomography By means of computer analysis, this technique is used in oceanic research to create images from data collected when acoustic signals are passed through an object. Ocean floor mapping programs make good use of this technology.

acrorhagi NEMATOCYST-armed defensive structures of sea anemones.

Actinopoda Phylum of the radiolarians and heliozoans, within the kingdom Protozoa.

actual evaporation The amount of water evaporated from a somehow limited water supply. *See* POTENTIAL EVAPORATION.

Aden, Gulf of A deep oceanic area centered at about 12° N and 47° E that separates the African coast, to the south, from the Arabian coast, to the north. It connects the RED SEA to the ARABIAN SEA.

Adriatic Sea An arm of the MEDITERRANEAN SEA centered at about 43° N and 16 E° that separates the eastern Italian coast from the Hellenic and Balkan Peninsulas. It is bounded by Italy to the west, Croatia to the east, and the IONIAN SEA to the south. Because of the high rate of freshwater feeding the Adriatic, its tidal range is three times greater than that of the rest of the Mediterranean.

Aegean Sea Centered at about 39° N and 25° E, this is a deep, marginal sea in the eastern MEDITERRANEAN SEA, strewn with volcanic islands. The Greek coast borders it to the west, Turkey to the east, and the islands of Crete and Rhodes to the south. Though largely tideless, a strait between Greece and the island of Evia has long suffered from unpredictable, sometimes violent tides. The ancient Greek philosopher ARISTOTLE commented on this anomaly.

AEPS *(Arctic Environmental Protection Strategy)* A program adopted in 1991 by the governments of the eight circumpolar nations (Canada, Denmark, Greenland, Finland, Iceland, Norway, Russia, Sweden, and the United States), whose objectives are to protect the arctic ecosystems, which includes protection of the arctic marine environment by taking steps to prevent marine pollution.

aerobic The state of requiring oxygen or living in oxygenated conditions.

aeroembolism *See* BENDS.

AFSC *(Alaska Fisheries and Science Center)* A part of NOAA (National Oceanic and Atmospheric Administration) that conducts marine research and fisheries stock assessment in the North Pacific Ocean and the Bering Sea.

See also Useful websites in the Appendixes.

aft A nautical term meaning located near the stern, or back, of a vessel.

agar Gelatinous product extracted from the walls of some red algae, commonly the oriental members of the genus *Gelidium.* Agar is used commercially in many applications, for example, as a solidifying agent in the preparation of candies, creams, lotions, canned fish, and meat. It is also used as an emulsifier in ice cream, as a clarifying agent in winemaking and brewing, and as a laboratory growth medium for microorganisms and in tissue culture.

agger *See* DOUBLE TIDE.

aggradation *See* ALLUVIATION.

agnatha One of two superclasses of fish, the Agnatha are the class of fish without jaws (such as lampreys and hagfish). This is opposed to the GNATHOSTOMATA, the class of fish with jaws (such as sharks, rays, and bony fishes).

Agulhas Basin An oceanic basin in the South Atlantic centered at about 43° S off the southern tip of Africa. Its deepest point averages roughly 19,100 feet (5,821 m).

Agulhas Current A part of the circulation of the southern Indian Ocean that moves southward along Africa's eastern coast. This current is one of the strongest in the world.

ahermatypic corals Non-reef-building corals.

Alaska Current A current of the North Pacific Ocean that enters the Gulf of Alaska heading north northeastward then turns to follow the land in a counterclockwise flow. It is the final branch of the ALEUTIAN CURRENT, strongest in the winter and weakest in July and August.

Alaska, Gulf of A wide body of water centered at about 58° N and 145° W. It is bordered by Alaska on all sides but the south, which is open to the North Pacific Ocean and therefore subject to all consequences of the weather. Despite its being battered by continual storms and occasional tsunamis, it is a region heavily fished.

Alboran Sea Part of the western MEDITERRANEAN SEA centered at about 4° W and 36° N. Morocco borders it to the south, Spain to the north, and it

GLOSSARY aeroembolism – Alboran Sea

connects to the North Atlantic by way of the Gibraltar Strait on its western end.

Aleutian Current A current of the North Pacific Ocean that sets eastward along the Aleutian Island chain and then splits; then one part flows into the BERING SEA and the other forms the ALASKA CURRENT.

Aleutian Trench A large deep-sea trench, skirting the southern border of Alaska's Aleutian Island chain and responsible for frequent earthquakes. Its deepest portion drops more than 24,000 feet (7,315 m). A 200-mile stretch, called the Shumagin Gap, is poised for a massive seismic disturbance. It is centered at about 51° N and intersects the international date line.

See also RING OF FIRE.

alevin A juvenile stage in fish, particularly in SALMON, in which the newly hatched baby is still attached to the yolk sac.

algae A diverse group of aquatic, plantlike organisms lacking adequate roots, stems, and leaves. They occur in many varieties, such as microscopic single cells; loose, hazy masses; knotted or branched colonies; and great seaweeds with rootlike holds and formations resembling stems and leaves. It has been estimated that algae account for more than 90 percent of the world's photosynthetic activity, making them the most important source of oxygen. Algae belong in the kingdom Plantae.

algae bloom The sudden overgrowth of algae due to abnormal changes that promote growth, such as pollutants that might raise the levels of organic matter, phosphorous, or nitrogen in the water. As the overgrowth dies it sets off abnormal bacterial growth that may become toxic and deplete the region of oxygen, resulting in the destruction of valuable oceanic biomasses.

See also EUTROPHIC; RED TIDE.

algal ridge A ridge of coralline algae on the perimeter of some CORAL REEFS.

algal turf A dense growth of algae.

Algerian Current A narrow, easily defined current of the Atlantic Ocean, setting eastward from the ALBORAN SEA into the MEDITERRANEAN SEA.

algology *See* PHYCOLOGY.

allopatric Pertaining to species or subspecies occurring in separate and mutually exclusive geographical regions. This is opposed to SYMPATRIC.

alluvial Pertains to accumulations of alluvium, which is SILT, SAND, or gravel, laid down by moving water. The fan-shaped layer deposited from a narrow valley stream onto a flood plain, for example, is called an alluvial fan or a DELTA.

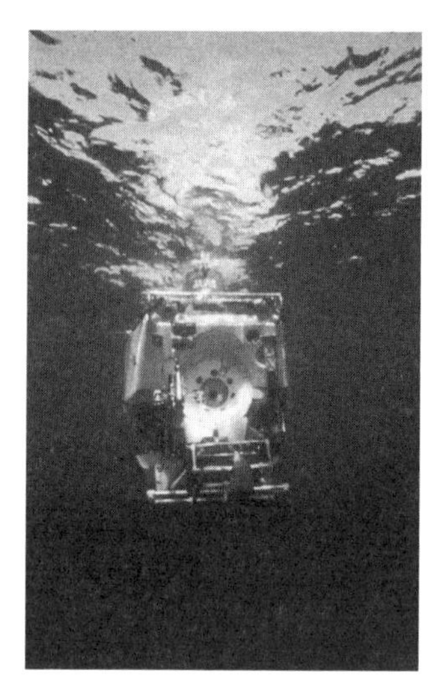

Alvin (Courtesy of NOAA)

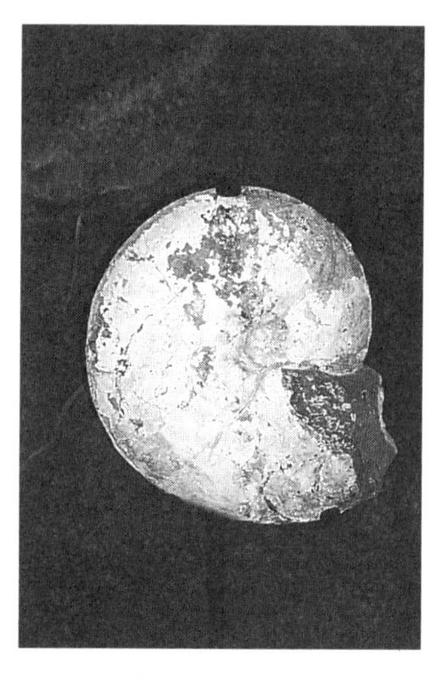

ammonite (Courtesy of Royal Tyrrell Museum/ Alberta Community Development)

alluviation The deposition of material carried by moving water. Also called aggradation.

alternation of generations A duel process of plant reproduction in which plant spores grow into a young plant—called a gametophyte—that produces a male and a female cell that unite to form a small, new plant—called a sporophyte—attached to the old plant. The sporophyte produces new spores that grow into a gametophyte again.
See also BRYOPHYTE.

Alvin Named both for the scientist Dr. Allyn Vine of Woods Hole Oceanographic Institution (WHOI), and Alvin the cartoon chipmunk, *Alvin* is a deep-sea, lightweight, staffed SUBMERSIBLE built in the United States in 1964, capable of descending to as much as 10,000 feet (3,048 m). It is operated by WHOI and has performed over a thousand deep-sea rescue missions and explorations, such as investigating undersea mountain ranges and assisting in the exploration of the wreck of the RMS *Titanic*.

amidships A nautical term meaning in or toward the center of the boat.

ammonites First referred to by the Roman naturalist PLINY THE ELDER (c. C.E. 23–79), as *ammonus cornua* (Latin for "horn of Ammon"), these NEKTONS are an extinct form of chambered, marine cephalopod related to the nautiloids and found today only in fossilized form. Many and varied species thrived in greatest abundance during the Mesozoic era. Ammonites are classified in the subclass Ammonoidea, class Cephalopoda, phylum Mollusca, kingdom Animalia.
See also FOSSIL.

Ampere Seamount Centered at about 35° N and 13° W, this is an undersea peak in the North Atlantic Ocean off the North African coast, northeast of the Madeira Islands. There are some who believe that stonework created by humans exists here and that it could be a location for the lost city of ATLANTIS.

amphibian Pertains to animals capable of living both on land and in water. These cold-blooded vertebrates, such as frogs, toads, and salamanders, are mainly freshwater animals.

amphidromic point A stationary point in the sea where the TIDAL RANGE is zero, or nearly so, and around which the peak of a standing wave or a high water level rotates once in each tidal cycle: counterclockwise in the Northern Hemisphere and clockwise in the Southern Hemisphere.

amphipod Flat, shrimplike, bottom-dwelling crustaceans and beach-dwelling sand fleas, typically colorless, with compound eyes not on stalks.

They are in the order Amphipoda below the phylum Arthropoda; the majority are marine, but a few species are freshwater.

ampullae of Lorenzini Pores forming an electrical receptor system, present in a number of cartilaginous fishes, that serve to detect weak electrical fields generated by other living creatures, even when they are hidden.

Amundsen Sea An arm of the Southern Ocean, it is a marginal sea named after the Norwegian explorer ROALD AMUNDSEN, which lies centered at about 73° S and 112° W, directly off the coast of Antarctica among the BELLINGSHAUSEN SEA, to the east; the ROSS SEA, to the west; and the imaginary Antarctic Circle to the north.

anadromous fish Refers to fish that spend most of their adult lives in the marine environment but breed in freshwater. SALMON are an example. Compare to CATADROMOUS and OCEANODROMOUS FISH.

Anadyr Current A surface current that flows along the northwest side of the BERING SEA and then continues through the BERING STRAIT. This current is a major factor for depositing biological material into Chirikov Basin, which lies north of Saint Lawrence Island, and thus the basin is an active feeding ground for whales.

Anadyr, Gulf of *(Anadyrskiy Zaliv)* A large embayment of the BERING SEA centered at about 64° N and 175° W, indenting the extreme eastern tip of Siberia.

anaerobic Existing in the absence of oxygen.

anal fin In fish, a ventral (belly) fin located behind the anus.

anatomy Pertains to the individual design or the scientific study of the structure of an entire organism.

Andaman Sea A body of water in the northeast end of the Indian Ocean centered at about 10° N and 95° E. The Andaman Islands border it to the west, the Malay Peninsula to the east, Sumatra to the south, and Burma's Irrawaddy Delta to the north. It connects to the JAVA SEA by way of the Malacca Strait.

anemometer An instrument for measuring direction and speed of wind. The most common anemometer works by a series of cups attached to horizontal limbs that spin around a vertical support. The rate of rotation is calibrated to obtain wind speed.

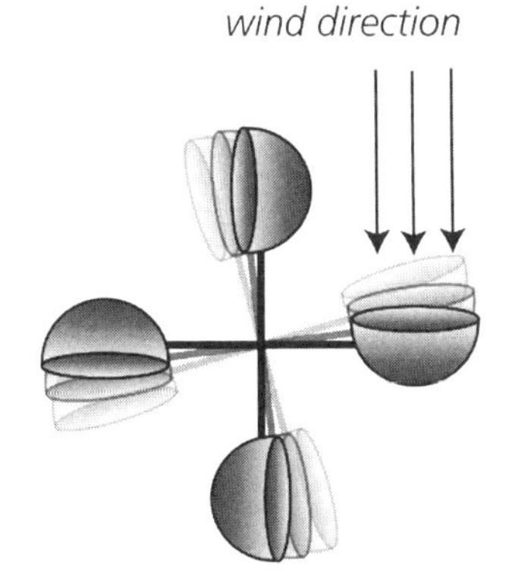

anemometer

angiosperm A plant whose seeds are produced in an ovary, such as a flower, fruit, or SEA GRASS, as opposed to a gymnosperm, whose seeds are not produced in an ovary, such as a conifer.

Angiospermae Phylum of the salt marsh grasses, sea grasses, and mangroves, within the kingdom Plantae.

Angola Basin A basin of the Atlantic Ocean located west of Africa directly along the prime meridian at about 15° S longitude. It lies between the Angola Abyssal Plain, to the south, and the Guinea Ridge, to the north. Also known as the Buchanan Deep.

Angola Current In the Atlantic Ocean, it is part of a subsurface gyre driven by the South Equatorial Countercurrent centered at about 13° S and 4° E at around 1,000 feet (304 m) in depth.

ANGUS *(acoustically navigated underwater survey system)* This equipment operated by SCRIPPS INSTITUTION OF OCEANOGRAPHY is essentially an unstaffed sled with observation cameras. In 1977 it made its first observation of deep-sea vent communities.

Animalia One of the five taxonomic kingdoms of living organisms whose members are multicellular, eukaryotic (having a true nucleus), and heterotrophic. The other kingdoms are Fungi, Monera, Plantae, and Protozoa. Organisms that ingest food from outside sources are classified as animals, as opposed to plants, which generate their own energy through PHOTOSYNTHESIS. Within this section, many marine-related phyla of the animal kingdom are recognized.

annelid Common name for wormlike invertebrates, such as the marine bristle worms, that make up the class Polychaeta, under the phylum Annelida. Recently, scientists discovered a new species of marine worm living in frozen methane in the deep waters of the Gulf of Mexico. New discoveries within the marine environment are being made each year.

Annelida Phylum of the polychaete worms and leeches, within the kingdom Animalia.

annulus The growth ring in a fish SCALE.

anoxic Deficient in oxygen.

Antarctic Cold climactic regions south of the Antarctic Circle, surrounding the South Pole. *Antarctic* means the opposite of *Arctic*.

Antarctic Circle Latitude line 66.5° south of the equator in Earth's Southern Hemisphere. That places it 23.5° north of the South Pole. On June 21 (winter solstice), the Sun never rises above the horizon, and on December 22 (summer solstice), the Sun never sets below the horizon. *See also* ARCTIC CIRCLE.

Antarctic Convergence An area of ocean encircling Antarctica between latitudes 47° S and 62° S that experiences rapid temperature changes from current subduction and convergence.

Antarctic Current The principal current of the Southern Ocean region, directly associated with the EAST WIND DRIFT and the WEST WIND DRIFT. The temperature and salinity effects of this powerful current are global.

Antarctic Ocean The name sometime applied to the SOUTHERN OCEAN, the fifth and most recently classified division of the WORLD OCEAN.

anterior The head or forward end of an animal.
See also POSTERIOR.

Anthozoa A class of marine COELENTERATES in the phylum Cnidaria (Coelenterata). The anthozoans are animals with radial symmetry, such as sea anemones and some corals.

anthracosaur A Paleozoic labyrinthodont amphibian. Labyrinthodonts typically resembled salamanders or crocodiles. They are considered to be the earliest TETRAPOD vertebrates.

Antilles Current A clockwise-circulating current of the North Atlantic Ocean. It is one of the distinct currents of this area's circulation system, which includes the NORTH EQUATORIAL CURRENT, GULF STREAM, North Atlantic Drift, and CANARY CURRENT.

AODC *(Australian Oceanographic Data Center)* Established within the Royal Australian Navy in 1964, this center participates in national and international projects in many branches of marine science.
See also Useful websites in the Appendixes.

AOML *(Atlantic Oceanographic and Meteorological Laboratory)* Part of the NOAA (National Oceanic and Atmospheric Administration) network located in Miami, Florida, whose mission is to conduct research programs in oceanography, tropical meteorology, atmospheric and oceanic chemistry, and ACOUSTICS.
See also Useful websites in the Appendixes; NOAA.

aphotic Devoid of light.

aphotic zone Deep, open-ocean region below the DISPHOTIC ZONE, sometimes called the midnight zone, where no light penetrates and where PHOTOSYNTHESIS may never occur.
See also EUPHOTIC ZONE.

aplacophora A wormlike MOLLUSK, rarely seen, having small spines that cover its body. *Aplacophora* is a Greek word meaning, "no shell."

apneustic breathing A breathing pattern in marine mammals whereby a number of rapid breaths alternate with a prolonged pause in breathing.

aquaculture Cultivation of aquatic organisms under controlled conditions.

Aqua-Lung Developed in 1943 by the French oceanographer JACQUES COUSTEAU and the French engineer Emile Gagnan, the Aqua-Lung is commonly used equipment today, called SCUBA gear. Collectively, it is a cylinder of compressed air connected by hoses to a facemask through a pressure-regulating valve, permitting a diver to stay underwater for many hours by providing an air supply.

Arabian Basin Centered at about 10° N and 65° E, this is an undersea region of the Indian Ocean, in particular, the southern part of the ARABIAN SEA. It is separated from the SOMALI BASIN (of the same sea) by the CARLSBERG RIDGE. Its deepest point is roughly 15,265 feet (4,652 m).

Arabian Sea A northwest arm of the Indian Ocean centered at around 16° N and 65° E, bounded by the Arabian Peninsula on the west, the subcontinent of India on the east, the greater Indian Ocean on the south, and Iran and Pakistan on the north.

Arafura Sea Centered at about 10° S and 137° E, this is a marginal sea belonging to the waters around northern Australia and Indonesia. Australia borders it to the south, Irian Jaya and New Guinea to the north-northeast, the TIMOR SEA to the west, and the Gulf of Carpentaria to the southeast.

Aral Sea Centered at around 45° N and 60° E, this is a saltwater lake resting along the border between Uzbekistan and Kazakhstan. It was once part of the CASPIAN SEA, before the last ice age.
See also TETHYS SEA.

archipelagic apron Common in the Central and South Pacific Ocean, a broad, fan-shaped or cone-shaped underwater slope(s) forming against a volcanic seamount(s). In typical development, a volcano erupts and depresses the surrounding crust; then the depression fills with lava and new sediment to form the archipelagic apron.

archipelago A group or chain of islands, an example of which is the Hawaiian Islands or Alaska's Aleutian Island chain.

Arctic Cold climactic regions north of the ARCTIC CIRCLE, surrounding the North Pole.

Arctic Circle Latitude line 66.5° north of the equator in Earth's Northern Hemisphere. That places it 23.5° south of the North Pole.
See also ANTARCTIC CIRCLE.

Arctic Ocean Ocean area north of North America and Eurasia. It is one of the five divisions of the WORLD OCEAN and is covered all winter with pack ice and therefore never totally navigable. Its surface area covers 5,427,000 square miles (14,056,000 km^2), one-third of that over continental shelf.

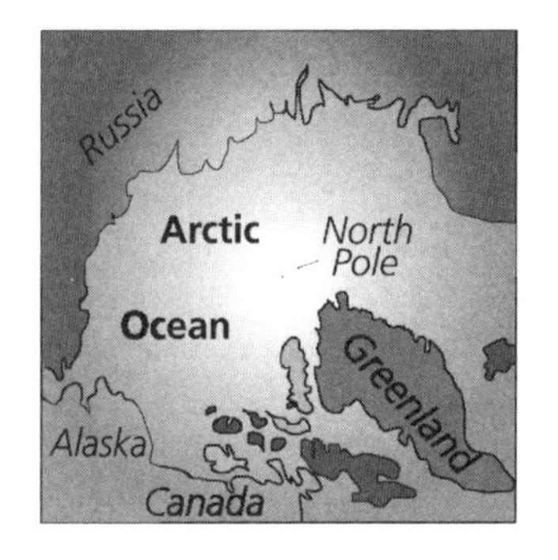

Arctic Ocean

GLOSSARY

Aqua-Lung – Arctic Ocean

See also ATLANTIC OCEAN, INDIAN OCEAN, PACIFIC OCEAN, *and* SOUTHERN OCEAN.

Argentine Basin A deep, cold-water basin of the South Atlantic Ocean centered around 45° S and 47° W off the coast of Argentina. Average depth is around 16,000 feet (4,876 m).

argonaut Ocean-dwelling mollusk related to the octopus that is also known as the paper nautilus. It has a fragile white or brown spiral shell, shaped rather like an angel wing, with many riblike folds. Argonauts belong to the genus *Argonauta,* family Argonautidae, order Octopoda, class Cephalopoda, phylum Mollusca, kingdom Animalia.

Aristotle's lantern In sea urchins, the five converging jaws and accompanying bony plates and muscles used to bite off food.

arrowworm Also called a chaetognath, a planktonic invertebrate characterized by an arrow-shaped, streamlined, transparent body with one or two pairs of lateral fins and a caudal fin that allows the animal to dart rapidly about. It grows from about 0.5 to four inches (1.3–10 cm) long. Arrowworms make up the phylum Chaetognatha, kingdom Animalia.

arthropod Invertebrate animal characterized by a segmented body and jointed appendages. CRUSTACEANS are arthropods, as are insects and spiders. Tallying in at around 875,000 known species, these animals make up the largest subdivision of the animal kingdom.

Arthropoda Phylum of the horseshoe crabs, sea spiders, crustaceans, insects, millipedes, and centipedes, within the kingdom Animalia.

artificial reef Foreign object or group of objects deliberately set on the bottom of a body of water within the EUPHOTIC ZONE with the purpose of attracting a variety of encrusting organisms, such as corals and sponges, to an area normally absent of significant quantities of sea life. This in turn attracts larger animals that feed on them, and then even larger animals are attracted, and so forth, until a complete food web has been created. At the time of this writing, the state of Alabama supports the largest artificial reef program in the United States, which includes some 1,200 square miles (3,108 km^2) of offshore waters.

artificial reef (Courtesy of NOAA/NURP)

aseismic ridge An undersea ridge not volcanically (seismically) active. Examples are the Arctic Ocean's LOMONOSOV RIDGE, which runs directly through the North Pole, and the South Atlantic's WALVIS RIDGE, a branch of the Mid-Atlantic Ridge that fingers up to the western African coast.

asexual reproduction Reproduction from the cells of a parent by development of an egg without fertilization (parthogenesis), division

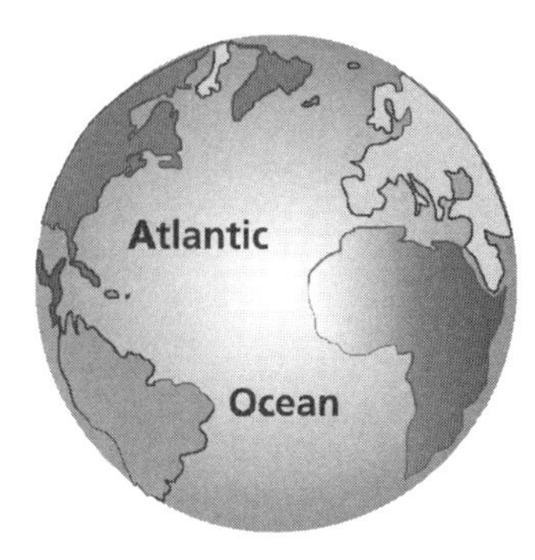

Atlantic Ocean

of a single complete organism into two complete organisms (fission), or a group of self-supporting cells that sprout then detach from the parent organism (budding).

ASLO *(American Society of Limnology and Oceanography)* A program whose goals are to promote interest in the study of freshwater and marine environments.
See also Useful websites in the Appendixes.

astern A nautical term meaning in the back of the ship. The opposite of *ahead.*

athwartships A nautical term meaning at right angles to the centerline of a boat. In a rowboat, the benches usually lie athwartships.

Atlantic-Indian Basin An oceanic basin centered at about 59° E and 55° S between the South Atlantic and South Indian Oceans. It lies southeast of southern Africa just west of the Kerguélen Islands near the coast of Antarctica and is bordered to the north by the Atlantic's CROZET BASIN. The greatest depths are roughly 15,530 feet (4,733 m).

Atlantic Ocean Earth's second largest of the five major oceans, second only to the Pacific and so heavily trafficked one might call it the Los Angeles Freeway of the waterways. Its surface area occupies about 29,640,000 square miles (76,767,600 km^2) and exists between the eastern shores of North and South America and the western shores of Europe and Africa.
See also ARCTIC OCEAN, INDIAN OCEAN, PACIFIC OCEAN, *and* SOUTHERN OCEAN.

Atlantic-type margin The trailing edge of a continent, as opposed to a PACIFIC-TYPE MARGIN, which is the leading edge of a continent.
See also CONTINENTAL DRIFT.

Atlantis A still-undiscovered island city of great mystery and legend, first mentioned by the Greek philosopher Plato in two of his dialogues, *Timaeus* and *Critias,* where he describes it as a utopian commonwealth ahead of its time. It was thought to be somewhere in the ATLANTIC OCEAN near the Pillars of Hercules (the ancient name for promontories flanking the east entrance to the Strait of Gibraltar). Most believe the accounts of Atlantis to be fiction, but the chances remain that records of the city, if a factual place, simply have not survived.

atmosphere *See* HYDROSPHERE.

atoll A ring-shaped CORAL REEF formed around a subsiding volcanic island, often one that has sunk into the sea leaving a central lagoon. It is one of three distinct types of coral reefs; the others are the BARRIER REEF and the FRINGING REEF.

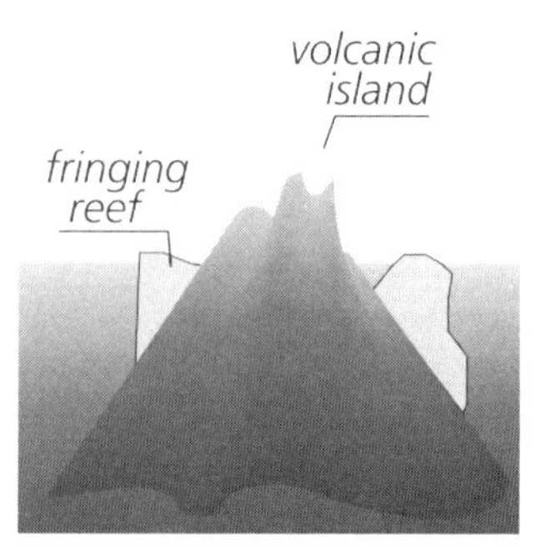

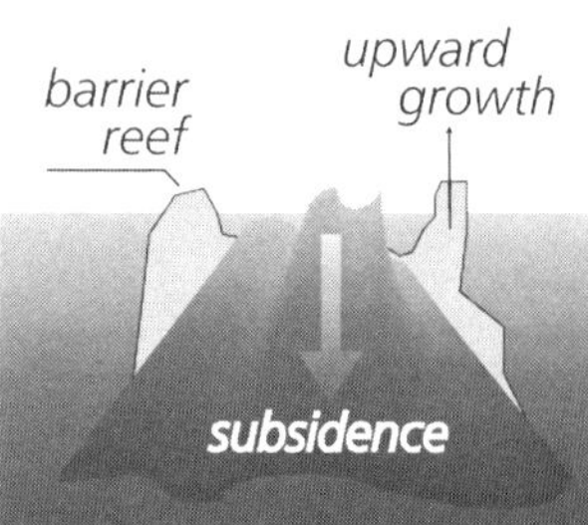

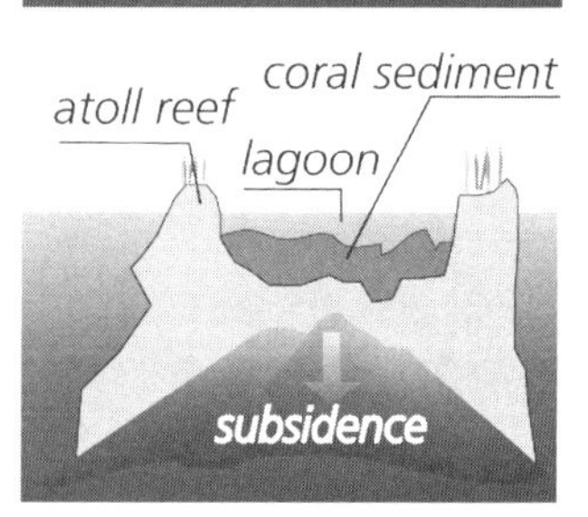

atoll

attenuation In the ocean water column, it is the decline of energy, such as light or sound, due to absorption and scattering.

authigenic One of three components of deep-sea marine sediment; the others are DETRITAL and BIOGENIC. Authigenic minerals are those formed by authigenesis: spontaneous crystallization within the water column or sediment during or after its deposition. (The authigenic minerals themselves make up only a fraction of sediment volume.)

autotrophs Described as "primary producers," these organisms can make organic materials from inorganic compounds, as in PHOTOSYNTHESIS, and thereby manufacture their own food.

AUV *(autonomous underwater vehicle)* An unstaffed robotic device that travels beneath the waves and can be equipped with technology such as SONAR or video cameras that perform desired functions, such as monitoring and surveillance for the U.S. Navy.

auxotrophic A species of phytoplankton unable to grow without certain compounds, which they are incapable of producing for themselves, such as vitamin B_{12}.

Aves The birds of Earth. A class in the subphylum Vertebrata, phylum Chordata, kingdom Animalia.

AWI *(Alfred Wegener Institute)* An institute in Bremerhaven, Germany, founded in 1980, that conducts scientific research about the polar regions, among other projects.
See also Useful websites in the Appendixes.

azimuth The horizontal angle between the observer's bearing on the Earth's surface and the line joining the observer and the object, measured clockwise conventionally from the north through east in astronomical calculations and from south through west in precise traverse and triangulation work.

azimuthal projections Also called zenithal projections or plane projections, whereby the round surface of the Earth is projected onto a flat map to demonstrate area and distance accurately, though some distortion in landmass shape can occur.

azoic zone A mid-19th-century term coined by SIR EDWARD FORBES pertaining to the deep part of the sea once thought (and wrongly so) to be not only devoid of life, but filled with thick, motionless seawater under tremendous pressure, therefore ensuring lifelessness. We have since discovered that this idea is inaccurate.

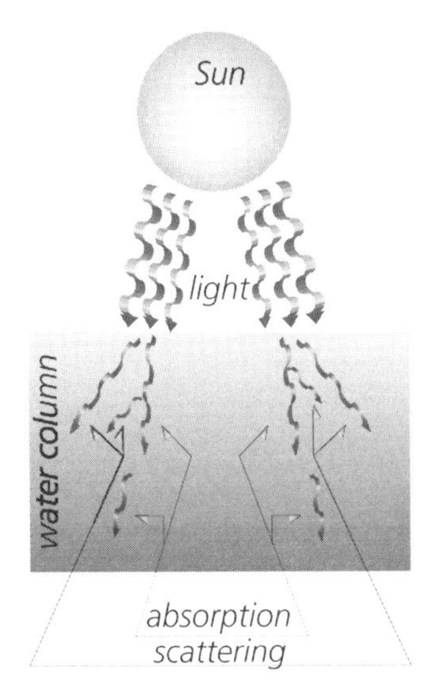

attenuation

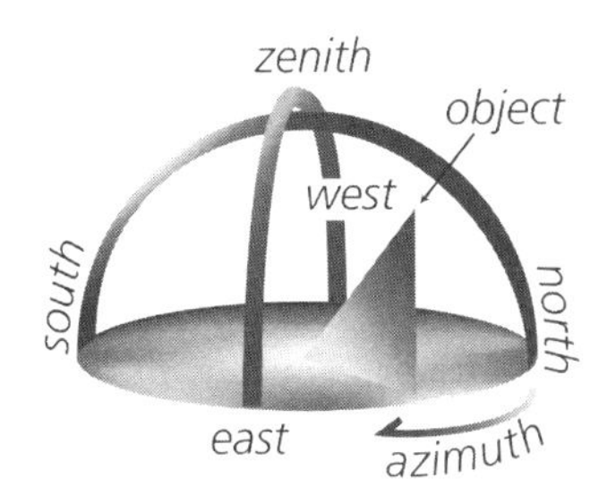

azimuth

Azores Current A branch of a North Atlantic Ocean subtropical gyre traveling approximately along 35° N–40° N that is an extension of the southeastern branch of the GULF STREAM. It splits into three main southward recirculation branches, which later join the westward-flowing NORTH EQUATORIAL CURRENT, and as such it is an essential part of the region's circulation.

Azov, Sea of A large, shallow gulf surrounded mostly by land and connected to the BLACK SEA by way of a narrow arm of water called the Kerch Strait. Centered at around 46° N and 37° E, it is bounded by Russia and Ukraine.

Bacillariophyceae Class of the DIATOMS, under the phylum Chrysophyta, within the kingdom Plantae.

back reef The inner region of an ATOLL or BARRIER REEF.

backshore zone The landwardmost section of the shoreline, sloping upward from the foreshore until interrupted by cliffs, dunes, or plant life such as a tree line. *See* NEARSHORE ZONE.

backwash In the marine environment, the water that races back into the sea after the breaking of a wave on the beach.

bacteria Single-celled microorganisms possessing no distinct cell nucleus, as opposed to fungi, plants, and animals, whose cells contain nuclei. Because they are the most common organism on Earth, one is not surprised to learn that they are connected intimately with all other living things. Only in colonies can bacteria be viewed by the naked eye. There are many different classifications of bacteria, but because of their uniqueness in relation to the rest of the world, they reside in a kingdom all their own, called Monera.

bacterioplankton Planktonic bacterium, the smallest plankton in the sea.

Baffin Basin Centered at about 72° N and 65° W, this is an oceanic basin residing between the Arctic and North Pacific Oceans within Baffin Bay. The point of greatest depth is roughly 7,870 feet (2,399 m).

Baffin Bay A cold northern body of water connected with the North Atlantic Ocean, to the south, through the DAVIS STRAIT and LABRADOR SEA and to the ARCTIC OCEAN, to the west, by way of various channels. Centered at about 72° N and 65° W, it lies between a battery of Canadian islands to the west and Greenland to the east and is covered most of the year by ice.

Balearic Sea Centered at about 40° N and 1.5° E, this is a body of water making up part of the northwestern MEDITERRANEAN SEA, lying directly between continental Spain and the Balearic Islands. The ALBORAN SEA lies to the southwest, the TYRRHENIAN SEA to the east.

baleen In nontoothed (Balaenoptera) whales, a horny slat that hangs crossways from the mouth with fringed inner edges used for filter feeding organisms, usually krill. Types of baleen whales are the blue, fin, humpback, minke, gray, and right whales.

Bali Sea A small sea, as seas go, belonging to the waters of Indonesia. It is nestled among the islands at about 7° S and 116° E, bounded by the FLORES SEA to the east and the JAVA SEA to the north.

Baltic Sea A marginal sea connected to the NORTH SEA by the waterways of SKAGERRAK and KATTEGAT. It is centered at about 57° N and 19° E, bounded by Sweden to the west, Poland to the south, the Gulfs of Bothnia and Finland to the north, and the republics of Lithuania, Latvia, and Estonia to the east.

Banda Sea A body of deep water between the Indian and Pacific Oceans and home to many volcanic islands of Indonesia. It is centered at about 5° S and 126° E, bounded by the TIMOR SEA to the south, the Ceram Sea to north, the ARAFURA SEA and New Guinea to the east, and the FLORES SEA and Sulawesi to the west.

bar *(in relation to measurement)* A unit of pressure commonly used by weather forecasters equal to about one atmosphere of pressure (0.987 atm.).

bar *(in relation to geography)* A sand or mud ridge lying across the mouth of a bay or river.

barbel Common name for a fish that possesses tactile (touch) feelers on the upper lips, which it uses to search for food.

Barents Sea Named for the Dutch explorer WILLIAM BARENTS, it is a body of water belonging to the ARCTIC OCEAN. Centered at about 75° N and 45° E, it lies between the Svalbard Archipelago, to the northwest; Novaya Zemlya Archipelago, to the east; and Asia, to the south.

barnacle Marine, shelled crustaceans (not mollusks) that, as adults, permanently attach themselves to other things (sessile), either living or nonliving. They range in length from less than one inch to 30 inches (2.5–76 cm) and feed by extending jointed legs, called cirri, that trap planktonic organisms. Barnacles constitute the subclass Cirripedia of the class Crustacea, phylum Arthropoda, kingdom Animalia.

barrier beach An extended sand or gravel ridge running parallel to but separate from the coastline, because of either a tidal salt marsh or a lagoon.

barrier island *See* ISLAND.

barrier layer In the water column, a mixed salinity layer separating the warmer HALOCLINE from the cooler THERMOCLINE.
See also PYCNOCLINE.

barrier reef One of three types of CORAL REEFS; the others are an ATOLL and a FRINGING REEF. A barrier reef is an offshore reef separated from land by a deep lagoon. The famous Great Barrier Reef, off Queensland, Australia, is the largest of its kind in the world, extending for about 1,250 miles (2,011 km).

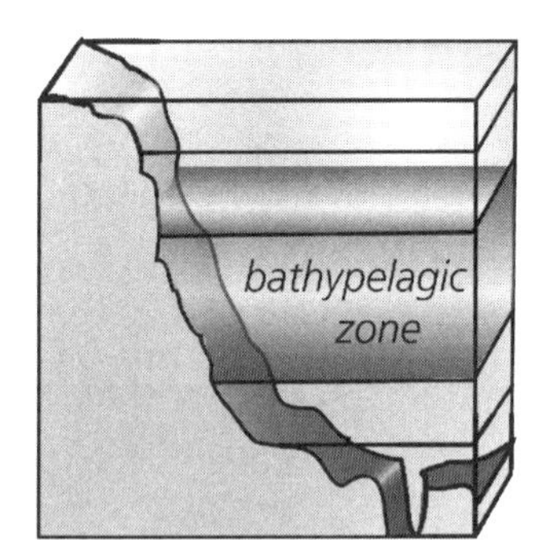

bathypelagic zone

basalt A dark volcanic rock found in the ocean crust. It is high in iron and magnesium.

Bass Strait A shallow waterway named for its discoverer, the English explorer GEORGE BASS, located at around 40° S and 146° E between Tasmania and southeastern Australia.

bathyl zone Marine ecologic region of seafloor between the edge of the continental shelf at about 300 to 1,000 feet (91–304 m) to the beginning of the abyssal zone at about 3,000 to 14,000 feet (914–4,267 m).

bathymetry Scientific measurement and charting of the ocean floor.

bathypelagic zone A zone of the open ocean beneath the MESOPELAGIC ZONE that runs between about 3,300 and 12,000 feet (1,006–3,657 m). Examples of life-forms found in this zone are dories, deep-water gulper eels, and scorpion fish.

bathythermograph
(Photograph by Captain Robert A. Pawlowski/ Courtesy of NOAA)

bathyscaphe A piloted, navigable diving craft (submarine) that can withstand the pressure of great depths, as far down as six miles! It consists of a heavily walled sphere, housing the crew and equipment, attached to a compartmental hull filled with lighter-than-water gasoline, which gives the craft its buoyancy. This is similar to yet distinguished from the earlier version called a bathysphere, which is a crewed, nonnavigable diving sphere suspended from a cable.

bathythermograph A scientific instrument developed by the British engineer Athelstan Spilhaus with the help of the geophysicist WILLIAM MAURICE EWING, both of whom were affiliated with Woods Hole Oceanographic Institute, which when towed behind a ship can make a temperature-versus-depth profile of successively deeper layers of the sea. The original mechanical bathythermograph (MBT) is obsolete except in a few countries; the more modern expendable bathythermograph (XBT) is the one more commonly used today.

bay A wide inlet of a sea or ocean, smaller than a gulf.

bay beach A coastal beach of a bay protected from wave and current erosion by headlands.

beach The coastal shore near a body of water between the high and low watermarks.

beach berm A sandy or shingled formation high on a beach caused by deposit of sediment by receding waves. A beach may have one or multiple berms.

beach face A sloping section of the beach, below the beach berm, exposed to wave SWASH.

beam A nautical term meaning the greatest width of a boat (usually amidships). If a boat is eight feet wide at its widest point, it has an eight-foot beam.

beam trawl In biological oceanography, a sampling device towed by a ship. A beam trawl is a net in a rectangular frame attached to towing cables designed to catch benthic organisms that live just above the bottom sediment.

bearing A nautical term meaning either the direction relative to the heading of a vessel or the direction of a vessel (or object) expressed as a true bearing as shown on a chart.

Beata Ridge Part of the seafloor of the CARIBBEAN SEA. It is centered at about 16° N and 72° W between the VENEZUELAN and COLUMBIAN BASINS, a region of unusually thick oceanic crust composed mainly of gabbros (granular igneous rock), dolerites (various coarse basalts), and rare pillow basalts.

Beaufort Sea Centered at about 73° N and 140° W, this is an iceberg-ridden body of water branching from the ARCTIC OCEAN. Alaska and Canada border it to the south and west, respectively, and the Arctic Ocean proper lies to the north.

Beaufort wind scale First designed in 1805 by the British naval officer SIR FRANCIS BEAUFORT, this scale uses the visible effects of wind to measure its force from 0, calm, to 12, hurricane.

belemnites Extinct class of invertebrate CEPHALOPODS sporting a molluskan foot modified into a ring of tentacles around the mouth.

Belle Isle, Strait of A commercially traveled waterway between the island of Newfoundland and mainland Canada that connects the Gulf of Saint Lawrence to the North Atlantic Ocean's LABRADOR SEA. It lies at about 51° N and 56° W.

Bellingshausen Abyssal Plain Centered at around 100° W along the ANTARCTIC CIRCLE, this is one of four oceanic plains encircling the continent of Antarctica; the others are the ENDERBY and WEDDELL PLAINS and the SOUTH INDIAN BASIN. The greatest point of depth is roughly 17,000 feet (5,181 m).

Bellingshausen Sea A marginal sea of the Southern Ocean lying between west Antarctica and the imaginary ANTARCTIC CIRCLE between 70° W and 100° W. It is named for its discoverer, the Russian admiral FABIAN GOTTLIEB VON BELLINGSHAUSEN.

bends Also called decompression sickness or aeroembolism, a painful, sometimes fatal condition that occurs when air pressure surrounding the body decreases too rapidly, as when a diver ascends (rises) quickly from the deep. Nitrogen escapes from the bloodstream and cannot rapidly reenter. The bubbles stretch and burst tissue and impair the circulation of blood, concentrating pain in the joints (hence the term *bends*) and potentially causing itching or tingling, breathing problems, and partial or total paralysis. The standard safe ascent rate for a scuba diver is no faster than 60 feet (18 m) per minute. The depth achieved along with the time spent at that depth determines one's need to spend time decompressing.

Bengal, Bay of Centered at around 15° N and 90° E, this is a body of water belonging to the northeast Indian Ocean. India borders it to the west and north, Thailand and Burma to the east, and the Indian Ocean to the south.

Benguela Current A wind-driven, cold-water bottom current of the South Atlantic Ocean that sets northward along the southwest coast of Africa between about 15° S and 35° S and is the eastern branch of the subtropical gyre.

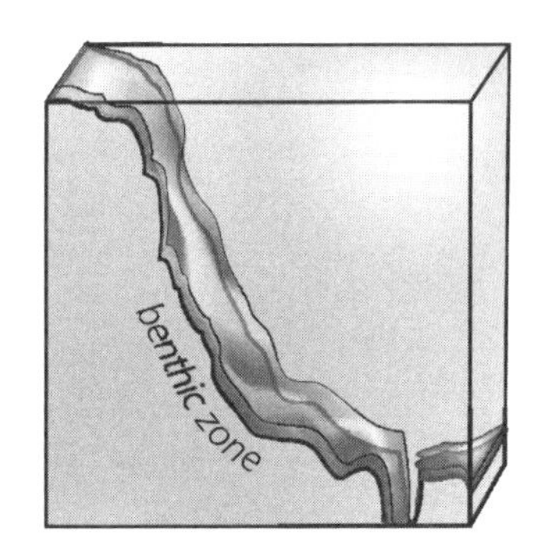

benthic zone

benthic zone The region on or near the bottom terrain of the ocean, regardless of the depth.

benthos Term used to describe creatures that permanently live on or around the bottom of the sea at whatever depth.

bergy bits Icebergs that have broken into pieces so small they officially are no longer considered icebergs and do not pose a threat to cruising vessels.

Bering Sea A cold, marginal sea of the North Pacific Ocean centered around 59° N and bisected by a zig of the international date line. Russia borders it to the west, Alaska to the east, the BERING STRAIT and thus the ARCTIC OCEAN to the north, and the Aleutian Islands to the south.

Bering Strait A narrow waterway centered at about 66° N along the international date line, separating Alaska from Russia and connecting the Arctic to the Pacific Ocean. As recently as 10,000 to 12,000 years ago during the last Ice Age this area, known as the Beringia land bridge, was exposed above sea level, allowing the migration of beasts and humans onto the continent of North America.

bight A wide bay formed by a bend in the shoreline. An example would be the Great Australian Bight in southern Australia.

bilateral symmetry In an animal, if split in two (down the axis), the halves would be identical.

bill The jaws of a fish or beak of a bird.

billfishes A fish group so named for their beaklike, or billlike noses, which includes MARLINS, spearfishes, and SAILFISHES.

binnacle On a ship, the stand that supports the steering compass.

biodiversity The variety of living organisms (plants, animals, microorganisms) within any given ecosystem, which is usually measured by the amount of different organisms found. An ecosystem that is biologically diverse is considered a healthy ecosystem.

biogenic One of three major components of deep-sea sediment; the others are DETRITAL and AUTHIGENIC. This sediment is chiefly made up of calcite and opal produced as the hard parts of organisms built up over generations. Calcite is formed by coccoliths (plants) and foraminifera (animals); opal derives from diatoms (plants) and radiolarians (animals).

biogenic reef An undersea ridge or rise composed mainly of biogenic matter.

biogeography Scientific study of the geographic distribution of living things.

biology 1) The scientific study of living organisms, their structure, function, growth, origin, evolution, and distribution, including botany and zoology. 2) The flora and fauna of a particular region.

bioluminescence Light produced by a living organism, referred to by scientists as cold light. Bioluminescent organisms glow when oxygen combines with a substance in the organism called luciferin in the presence of the enzyme luciferase. Some species of worms, fish, bacteria, crustaceans, fungi, jellyfish, mollusks, sponges, and protozoa can emit light. For the most part, this phenomenon is not fully understood and may function as a form of communication.

biomass The total of animal and plant organisms in a specific area at a specific time; also applied in calculating the abundance of a specific species.

biosphere *See* HYDROSPHERE.

biota All living things, plants and animals, of a region. When people talk about the biota of this or that area, they are speaking of the sum of diversity of living organisms in that area.

biotic Factors that pertain to life.

biotic crisis When the extinction rate outpaces the reproduction rate, causing mass extinction.

bioturbation The stirring up of bottom sediment by the movements of animals.

biped Descriptive of a two-footed animal.

Biscay, Bay of Part of the eastern ATLANTIC OCEAN, it is a body of water subject to heavy weather and known for its strong currents. It is centered at about 46° N and 4° W and is bounded by Spain to the south, the Atlantic Ocean to the west, France to the east, and the United Kingdom to the north.

bivalve MOLLUSK in the class Bivalvia under the phylum Mollusca, having two hinged shells. Examples of bivalves are clams, cockles, mussels, oysters, scallops, and shipworms.

black coral Colonial anthozoan that secretes a black protein skeleton.

Black Sea The world's largest inland water basin, connected to a small sister bay, the Sea of AZOV, to the north, and the MEDITERRANEAN SEA by way of the tiny twiglet of water called the Bosporus to the southwest. It is slightly saline within the top layers, growing denser and more saline with depth. Centered at about 43° N and 35° E, it is bordered by Turkey to the south, Russia and Ukraine to the north, and Romania and Bulgaria to the west.

black smoker *See* HYDROTHERMAL VENTS.

black swallower A deep-sea dweller of the perch family, capable of swallowing fish as large as itself with its loosely hinged jaws and curved, needlelike teeth. Within the water column, it exists from the lower MESOPELAGIC ZONE well into the BATHYPELAGIC ZONE.

blade Flat, photosynthetic "leafy" part of an alga or seaweed.

blenny A small marine fish of which exists an abundance of different species found all around the world. It is characterized as having an elongated, tapering body and a dorsal fin running its length. Some varieties use their two ventral fins to wriggle about the rocks out of water. These spiny-rayed fishes thrive in salt marshes and shallow reefs, feeding on small crustaceans. They can grow up to three feet (91 cm) in length and are good as food fishes, sold commercially as "ocean catfish." Striped blennies live in the Tropics and freckled blennies live farther south. Blennies belong to the families Blenniidae and Nototheniidae, order Perciformes, class Osteichthyes, subphylum Vertebrata, phylum Chordata, kingdom Animalia.

blow The sound and cloud of vapor characteristic of the exhalation of cetaceans when they surface to renew their air supply.

blowhole The nasal opening(s) on the top of a cetacean's head from which the blow arises; in toothed whales, the opening is single, and in baleen whales there is a double set.

blubber In marine mammals, this is the layer of fat lying just below the skin that helps store energy and preserve body heat as well as aid buoyancy. Blubber can be an important source of oil for lamps, soap, candles, and lubricants.

bluefish A most bloodthirsty ocean fish, roaming in schools through warm waters, preying in frenzies even after hunger is satisfied, and sometimes leaving blood trails in the current miles long. In 1935 masses of bluefish off Cape Hatteras, North Carolina, littered miles of beach with fright-stricken menhaden and silversides and actually grounded themselves on the shore in wild pursuit. They are popular with sport fishers and serve as an important food fish along the Atlantic seaboard. Bluefish belong in the family Pomatomidae, order Perciformes, class Osteichthyes, subphylum Vertebrata, phylum Chordata, kingdom Animalia.

blue holes Standing out as dark blue sinkholes visible in shallow NERITIC ZONES, these are underwater geographic features caused by water action that dissolves soft limestone deposits and most likely occurred during the last Ice Age, when water levels were lower. Examples of blue holes can be seen in the Caribbean's Grand Bahamas Bank.

BODC *(British Oceanographic Data Center)* A world-class data center supporting marine science in the United Kingdom.
See also Useful websites in the Appendixes.

body whorl The last and typically largest coil (whorl) of a GASTROPOD shell.

Bohol Sea A small sea centered at about 9° N and 124° E between the Pacific and Indian Oceans amid the southern Philippine Islands. The island of Mindanao lies to the south and Negros, Bohol, and Leyte to the north.

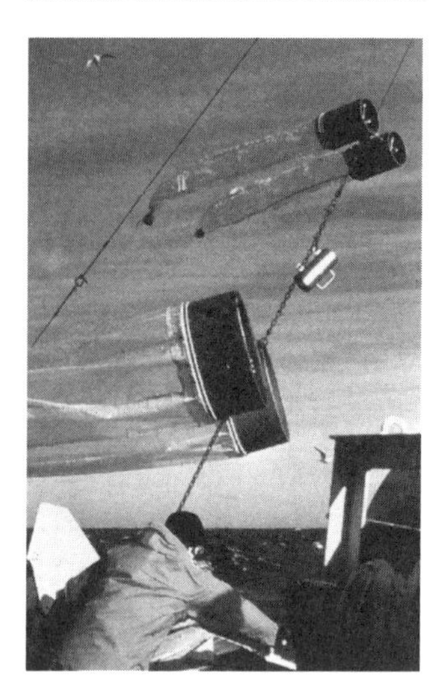

bongo net (Photograph by Commander Gerald B. Mills/Courtesy of NOAA)

bollard A heavy post at the edge of a pier or wharf to which a ship's lines may be fastened.

bongo net A twin set of nets, resembling bongo drums, that serve as a zooplankton sampling device. It can be towed from a vessel either vertically or diagonally from a selected depth to the surface. These types of experiments provide scientists with an idea of the kind of plankton that live within a particular depth range.

boom In nautical terms, the spar (support) used to extend the foot of a fore-and-aft sail.

bore *See* TIDAL BORE.

borers Marine organisms that specialize in boring holes into wood, rock, other organisms, or other materials, not to feed, but to make a cozy home from which they can casually filter feed while hiding. Examples of borers are the shipworm, gribble, and boring sponges. Borers can make a nuisance of themselves by destroying not only ship hulls but also the pilings of docks, marine warehouses, wharves, and other seaside structures.

boring sponges Sponges that bore through calcareous skeletons and shells, important because they enable the calcareous material to be transformed into a solution (liquid form) and reused to build new constructions. Boring sponges belong to the silica sponges under the phylum Porifera, kingdom Animalia.

botany 1) The scientific study of plants. 2) Plant life of a specific area.

Botany Bay A small inlet along the southeastern coast of Australia, important as the first landing place of the English navigator CAPTAIN JAMES COOK in 1770. It is located at about 151° E and 34° S.

Bothnia, Gulf of Northern arm of the BALTIC SEA centered at about 62° N and 20° E. Sweden bounds it to the west, Finland to the east, and the Baltic Sea proper to the south.

bottom currents Distinct, directional water movement at the ocean bottom, completely distinct from the movement of surface water. Temperature differences and the mere spin (rotation) of the Earth are two examples of the cause of their existence; ripple patterns in bottom sediment and rocky bottoms swept clear of loose material hold the proof.

bottom water Deep water moving toward the equator after running off from the polar regions, characterized by a higher oxygen content and a lower salinity than the waters through which it moves. Eventually, of course, it mixes with the ocean at large and therefore can no longer be designated as bottom water.

bow A nautical term for the front section of a vessel.

bowsprit A large, strong SPAR standing horizontally from the bow of a vessel.

brachiopod *See* LAMPSHELL.

Brachiopoda Phylum of the lampshells, within the kingdom Animalia.

brackish water Unpalatable mix of seawater and freshwater in no particular proportion.

branchial Pertaining to the gills of aquatic animals.

Brazil Basin An undersea basin of the South Atlantic Ocean located off the eastern coast of Brazil and centered at about 15° S and 25° W. The point of greatest depth is roughly 19,700 feet (6,004 m).

Brazil Current A comparatively weak current of the South Atlantic Ocean moving southwestward along the central coast of South America. It forms the western limb of the South Atlantic's subtropical gyre.

breaching Characteristic of a whale leaping either partially or fully from the water, followed by a loud splash as a spray of water erupts from the sea. Compare to SPYHOPPING.

breaker A wave that breaks into foam when it nears the shore.

breaker zone Area of the NEARSHORE ZONE where incoming waves from the offshore zone lose their stability and "break."

breakwater Either a natural or a constructed mound of rock rubble built up parallel to the shore whose presence (or purpose, when built) creates an area of safe, calm water where destructive wave action would otherwise cause damage.

brig Also called brigantine, this is a square-rigged sailing vessel sporting two masts. A hermaphrodite brig is rigged on the foremast as a brig and on the mainmast as a SCHOONER.

brine Said of the water of the sea. Brine is simply salty water but more specifically is saturated with sodium chloride.

Bristol Bay Centered at about 58° N and 160° W, this is a large embayment of the BERING SEA and home to the largest sockeye salmon runs in the world. The broad arm of Alaska's Aleutian Island chain cradles its southern and eastern boundaries.

British Royal Society *See* ROYAL SOCIETY OF LONDON.

breach

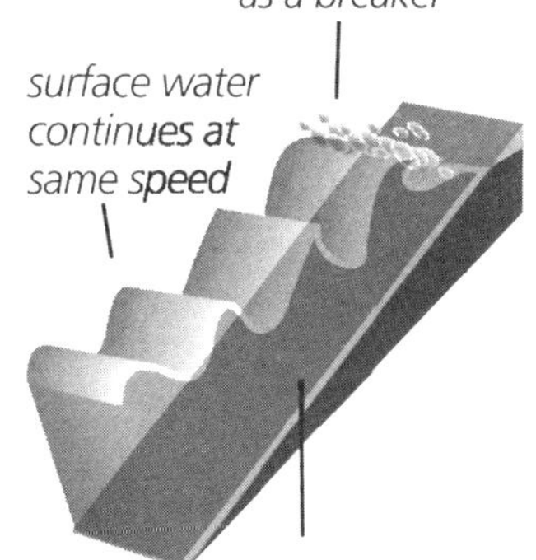

breaker

brittle stars Echinoderms, closely related to starfish and grouped with the basket star. They have five flexible arms, readily broken off, that radiate from a central disk and tube feet that are used in feeding. The group includes about 2,000 species. Brittle stars belong in the class Ophiuroidea, phylum Echinodermata, kingdom Animalia.

broach A nautical circumstance in which a ship has turned parallel to the waves and is in danger of capsizing, or rolling over.

broadcast spawner Marine animal that reproduces by releasing its eggs and sperm into the water, allowing external fertilization to occur.

bryophyte Also called Bryophyta, the plant phylum of some embryo-forming plants, which includes the mosses, hornworts, and liverworts. These nonvascular, spore-producing plants grow in moist places, absorbing water directly from the base on which they grow or from the air.

Bryozoa Phylum of colonial marine invertebrates, within the kingdom Animalia.

bryozoan Also called moss animal, this is a sessile, colonial, aquatic invertebrate whose variable species number around 5,000 worldwide. The marine species may resemble seaweed or may appear as lacy, hard-shelled, or mosslike crusts or mounds on solid surfaces, such as rocks or pilings. There are three classes of bryozoans: Gymnolaemata (primarily marine animals), Phylactolaemata (freshwater animals), and Stenolaemata (marine animals).

See also COLONIAL ORGANISMS; ZOOID.

bucket temperature The ocean's surface temperature when measured with a bucket thermometer or when measured by immersion of a special surface thermometer into a newly drawn bucket of water.

bucket thermometer A special thermometer used to measure ocean water temperatures by lowering it into the sea in a bucket and, after a time, hauling it up with the surrounding water acting as an insulator, then accurately reading the results.

bulkhead On a ship, the vertical partition that acts as a hull support and separates compartments.

buoy A floating metal or wooden object typically moored within dangerous boating areas, marking a mooring place or submerged crab pot or marking a channel for safe passage. Red buoys mark the landward side, green the seaward (in case of fog). Some buoys not only serve as a visual aid to ships but also record environmental information such as wave height and wind velocity.

buoy (Courtesy of NOAA)

buoyancy The ability of an object to float in water or to rise if it is less dense than the liquid within which it resides.

buoy tender Any government-run vessel whose purpose is tending to the operational capability and proper deployment in the sea of navigational or research buoys.

buttress Seaward face of a coral reef, extending from a depth of about 70 feet (21 m) to just below the low tide line.

cabbeling In oceanography, the mixing of two bodies of water of different temperatures and salinity but the same density to form a third body of water having a greater density than either of the two original constituents.

Cabot Strait An arm of water connecting the Atlantic Ocean with Canada's Gulf of Saint Lawrence, lying between Nova Scotia and Newfoundland. It is centered at about 47° N and 59° W.

caisson Enclosed on all sides but the top, which receives the intended piling, a box-shaped foundation that is sunk into river or ocean bottoms to support pilings for structures such as bridges or piers.

calcareous Pertains to things made of CALCIUM CARBONATE.

calcareous ooze A type of biogenic sediment made of CALCIUM CARBONATE shells and skeletons of marine organisms.

calcichordate An extinct aquatic burrower from the Cambrian to Devonian period, often grouped with echinoderms, with an exoskeleton formed of plates. Some authorities believe they hold a pivotal position in the evolution of early vertebrates.

calcium carbonate A white, crystalline, nonsiliceous chemical compound that occurs as a natural mineral and is also a major component of the shell, skeleton, and various other parts of many organisms. In water, it typically is insoluble unless the water contains dissolved carbon dioxide.

See also FOSSIL.

California Current A current of the North Pacific Ocean driven by the wind in a southeasterly direction along the west coast of the United States and Baja California until it joins the NORTH EQUATORIAL CURRENT. It is a region of upwelling and an eastern branch of the North Pacific's clockwise flowing subtropical gyre.

California, Gulf of Centered at about 28° N and 112° W, this is an inlet of the Pacific Ocean identified on some maps as the Golfo de California. It separates the distinctive Baja Peninsula from Mexico.

Canada Basin One of four major basins of the Arctic seafloor beneath the ARCTIC OCEAN; the others are the MARKOV, FRAM, and NANSEN BASINS. It is centered at about 78° N and 145° W and its greatest point of depth lies roughly at 12,000 feet (3,657 m).

Canary Basin A basin of the North Atlantic Ocean so named for its proximity to the Canary Islands directly to its east. It is centered at about 27° N and 25° W and averages roughly 16,000 feet (4,877 m) in depth.

Canary Current A cool, fog-producing current of the North Atlantic Ocean moving southward off the west coast of Portugal and along the northwest coast of Africa. It is a branch of the NORTH ATLANTIC CURRENT.

cape 1) A region of dark pigmentation on the backs of some dolphins around the dorsal fin. 2) A pointy landmass that juts into the sea.

Cape Verde Basin A large basin in the North Atlantic Ocean centered at about 18° N and 33° W between the northwest coast of Africa and the undersea Mid-Atlantic Ridge. It is part of the Cape Verde Abyssal Plain and ranges between 7,000 and 17,500 feet (2,133–5,334 m) in depth.

captacula Feeding tentacles that extend from the head of some mollusks.

carapace The term used when describing the hard covering of animals such as the shell of a turtle (or tortoise) or the hard plate covering the head and thorax of an arthropod.

caravel A 15th-century sailing vessel, typically Spanish or Portuguese, rigged with three or four masts, all with lateen (triangular) sails except the foremast, which was rigged with a square sail. Its shape was a broad bow and a high, narrow poop deck. The ships used by CHRISTOPHER COLUMBUS on his voyages of discovery were all caravels.

carbon An element on which all known life on Earth is based, in company with hydrogen, oxygen, phosphorus, and nitrogen. Carbon forms large organic molecules that include amino acids, enzymes, sugars, and other chemicals vital to life.

carbon cycle The continuous circulation of carbon, in whatever form, through the Earth's ecosystem. Animals exhale carbon dioxide (CO_2) as an atmospheric gas as a metabolic by-product. Plants and algae use the gas for PHOTOSYNTHESIS. Animals and fish eat the carbon-based plants and algae, ingesting the carbon, which again is released as a by-product. CO_2 is also produced by animal waste decomposed by bacteria, volcanic eruption, and the burning of fossil fuels.

GLOSSARY Canada Basin – carbon cycle

Caribbean Current A clockwise-moving, warm-water current of the CARIBBEAN SEA setting westward past Mexico's Yucatán Peninsula and Cuba, where it helps form the FLORIDA CURRENT and ultimately the GULF STREAM.

Caribbean Sea A marginal sea of the Atlantic Ocean centered at about 15° N and 78° W. Mexico borders it to the west, Colombia and Venezuela to the south, the Greater Antilles to the north, and the Lesser Antilles and Atlantic Ocean to the east.

Carlsberg Ridge An undersea ridge in the North Indian Ocean that begins at the Owen Fracture Zone and goes on to develop the Mid-Indian Ridge, which eventually merges into the Southeast Indian Rise. It is centered at about 5° N and 63° E off Africa's northeastern coast.

carnivore An animal that preys exclusively on other animals and eats their meat.

Carpentaria, Gulf of Arm of the ARAFURA SEA centered at about 14° S and 139° E between Australia's characteristic "dog ears" on its northern coast. In 1606 the Dutch explorer William Jansz first explored the gulf.

carrack A European merchant ship of the late Middle Ages, strongly built, similar to but larger than a CARAVEL. It had large castles fore and aft that never failed to catch additional wind, wanted or not. Of its three masts, two—the fore and main—were rigged with square sails, whereas the mizzen (aft mast) was rigged with a lateen (triangular) sail.

cartilage A tough, fibrous, elastic connective tissue that helps support vertebrate skeletons.

cartilaginous fish Also called Chondrichthyes, these fish are characterized by their flexible cartilaginous skeletons. Unlike other fish with bony skeletons, they possess no swim bladder to keep them afloat and, therefore, unless they are sedentary bottom-dwellers, must always stay on the move or otherwise sink. Examples of cartilaginous fish are sharks, skates, rays, and chimeras, which form the class Chondrichthyes, phylum Chordata, kingdom Animalia.

cartography The science, art, techniques, and production of mapmaking.

Caspian Sea Famous as the largest lake in the world, it is centered at about 42° N and 51° W, bounded by Iran to the south, Russia to the west, the republic of Kazakhstan to the north, and the republic of Turkmenistan to the east. The Caspian has less input than evaporation and is therefore saltier than other saltwater lakes.

See also ARAL SEA; TETHYS SEA.

catadromous fish Fish that live all of their adult life in freshwater and migrate to the sea to spawn. Freshwater American eels are an

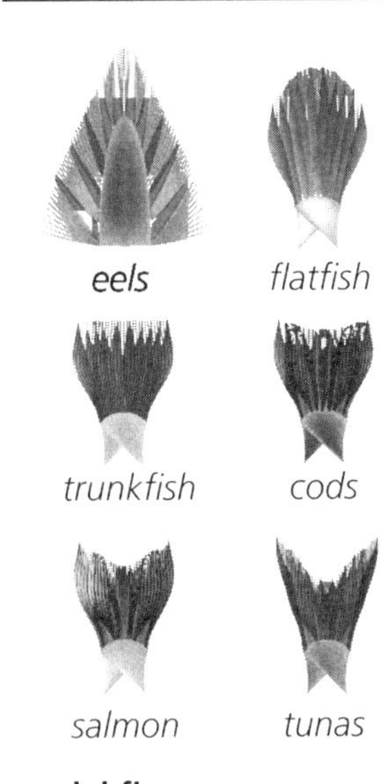

caudal fin

example of catadromous animals. Compare ANADROMOUS; OCEANODROMOUS.

catamaran A twin-hulled ship whose hulls are parallel, a design that is very stable in water. A catamaran can be either a sailing or a motorized vessel.

caudal fin In fish, it pertains to the tail fin. The shape of the tail fin can help in identifying the species of a fish.

caudal peduncle In fish, the area where the CAUDAL FIN joins the rest of the body. *See also* TAIL STOCK.

Ceara Abyssal Plain An oceanic plain of the mid-Atlantic Ocean centered at about 38° W along the equator. It separates South America's northern continental shelf from the Mid-Atlantic Ridge; its point of greatest depth is roughly 12,600 feet (3,840 m).

Celebes Sea Centered at about 3° N and 122° E, this is a body of water contained between the Indonesian and Philippine Islands, separating Sulawesi, to the south, from the Philippines, to the north.

Celtic Sea A shallow sea of the North Atlantic Ocean centered at about 51° N and 7° W. England borders it to the east, Ireland and Saint George's Channel to the north, and the Atlantic Ocean to the south and west.

central nervous system The complex of vertebrate nerve tissues consisting of the brain and spinal cord that control the body's activity.

Central Pacific Basin A basin of the Central Pacific Ocean centered at about 5° N and 175° W between the Marshall Islands to the west and the Hawaiian Islands to the east. The point of greatest depth is roughly 16,050 feet (4,892 m).

central rift valley A depression in a SPREADING RIDGE.

cephalopod Refers to an animal that has tentacles attached to its head, such as the cuttlefish, squid, and octopus.

cephalothorax The fused head and thorax of an animal.

cerata Fingerlike projections along the dorsal sides of some NUDIBRANCHS.

ceratotrichia The horny, filamentous rods that peripherally surround the large, thick fins of cartilaginous fishes.

cerebroganglion The ganglion that makes up the brain of an invertebrate. (plural, *cerebroganglia*)

cetacean Marine mammal, including all WHALES, DOLPHINS, and PORPOISES, characterized by a nearly hairless body, broad flippers, and flat, notched tail. Cetaceans make up the order Cetacea, class Mammalia, phylum Chordata, kingdom Animalia.

chaetognath Free-swimming, carnivorous, pelagic, wormlike planktonic animal.

Chaetognatha Phylum of the arrowworms, within the kingdom Animalia.

channel 1) The bed of any watercourse and its sides. 2) The deep navigable passage of a river or harbor. 3) A broad strait linking two seas.

chart The correct term to use when referring to a "map" of oceans or waterways for use in navigation and to other nautically related data. This is opposed to the term *map,* which when correctly used pertains to continental information and guidance.

chelipeds The first pair of pincerlike legs in most decapod crustaceans, used for seizing and crushing prey.

chemoreceptor A sensory organ in animals that responds to contact with chemical substances. Chemoreceptors exist abundantly in a cephalopod's tentacles or in the head of certain marine worms. *See also* MECHANORECEPTOR.

chemosynthesis Organic compounds formed through energy derived from the inorganic chemical reactions of materials such as hydrogen, ammonia, and sulfur.

chemotroph Phytoplankton that can flourish without any light. Organisms that are facultative (having the option) can either take it or leave it, and species that are obligate (possessing restrictive needs) are completely unable to live without it.

chert A fine-grained sedimentary rock typically made of millions of globular siliceous skeletons of tiny marine plankton called RADIOLARIANS. Flint is black chert.

Chesapeake Bay Centered at about 38° N and 76° W, this is the Atlantic Ocean's largest bay along the North American coast. It separates Virginia from Maryland and is an outlet for many rivers such as the Potomac, Patuxent, and Susquehanna.

chimaera Cartilaginous fish related to sharks. An example is the ratfish, which lives in temperate waters; has large, sharklike eyes and a long tail; and can reach up to six feet (1.8 m) in length. Many species possess a poison spine positioned before the dorsal fin. Chimaeras belong to the subclass Holocephali, class Chondrichthyes, subphylum Vertebrata, phylum Chordata, kingdom Animalia.

China Coastal Current A current of low salinity that moves southward along the Chinese coast in the YELLOW SEA. It flows into the East China Sea, where it divides; part of the current continues down the coast and part flows into the Taiwan Current.

chine line On a flat or V-bottomed boat, the intersection that joins the bottom and sides.

chitons Invertebrate BENTHOS, all 700 or so species of which are marine. Eight overlapping plates characterize their shells, which have bilateral symmetry. They are flat animals with the appearance of oval worms that skulk along rocky shorelines and benthic areas, using a broad, flat foot in search of seaweed and algae to eat. Chitons make up the class Polyplacophora, under the phylum Mollusca, kingdom Animalia.

chloride ion The most abundant dissolved chemical ion in SEAWATER, about 54 percent of the total weight of all dissolved inorganic substances. A chloride ion combined with a sodium ion composes common salt ($NaCl$).

chlorophyll The pigment found in plants that gives them their green color by absorbing red, violet, and blue light and reflecting only the green. Chlorophyll plays a basic role in PHOTOSYNTHESIS by taking light energy and converting into chemical energy.

Chlorophyta Phylum of the green algae, within the kingdom Plantae.

Choanoflagellida Phylum of the solitary and colonial zooflagellates, within the kingdom Protozoa.

chock and cleat Pertaining to the ever-present need for making fast a ship's lines, a chock is a typically U-shaped fitting through which a line is led, and therefore reducing chafe; a cleat is a classically anvil-shaped fitting to which the lines are made secure.

chondrichthyes *See* CARTILAGINOUS FISH.

Chordata Phylum of the amphibians, reptiles, birds, mammals, sharks, rays, sea squirts, tunicates, lancelets, vertebrates, lampreys, hagfish, chimaeras, and bony fishes. These fall within the kingdom Animalia.

chordates Vertebrate organisms characterized by having a hollow dorsal nerve cord, gill slits, and a notochord (flexible supporting axis of the body).

chromatophores A cell or group of pigment cells that can be controlled by an animal's nervous system to produce a new tone or color by changing its shape. This enables the animal to blend with almost any terrain, if it so chooses. Examples of animals possessing chromatophores are the cephalopods, such as the cuttlefish and squid, which are masters at color, pattern, and texture changes.

GLOSSARY
China Coastal Current – chromatophores

chronometer A precise timepiece.
See also HARRISON, JOHN.

Chrysophyta Phylum of the diatoms, within the kingdom Plantae.

Chukchi Sea A shallow sea of the ARCTIC OCEAN centered at about 70° N along the international date line. Both Russia and the East Siberian Sea border it to the west, Alaska to the east, the Bering Strait to the south, and the Arctic Ocean proper to the north.

ciguatera Name given to a type of poisoning caused by tropical fishes. The poison itself is called ciguatoxin and may be produced by DINOFLAGELLATES. Humans can contract ciguatera by eating seafoods contaminated with ciguatoxin.

Ciliophora Phylum of the ciliated protozoans, including tintinnids. These fall within the kingdom Protozoa.

cilium Hairlike structure used by animals for feeding or locomotion (plural, *cilia*).

circalittoral zone Another name for the outer INFRALITTORAL ZONE.

circulation The continual surface and/or deep-water movement of the oceans.

clade A group of organisms believed to have evolved from a common ancestor.

cladogenesis The splitting off of an evolutionary lineage into one or more separate lineages. The systematic classification of these groups is called cladistics.

clam A marine invertebrate species of bivalve mollusk found burrowed in saturated beach sand, existing as inconspicuous filter feeders. Bold evidence as to their buried location is the telltale squirt of water shooting heavenward from the sand as they pull their siphons back into their shells. This activity never fails to generate musical giggles and a one-sided game of hide-and-go-seek from children playing on the beach. Clams belong to the class Bivalvia, phylum Mollusca, kingdom Animalia.

class In TAXONOMY, a classification group consisting of closely related genera, below a phylum and above an order.
See also KINGDOM.

clastic Sediment or beach deposits made of rock fragments that have been transported a great distance from their original location.

clay Particles less than 0.002 inch in diameter. Typically they are hydrated aluminum silicates, defined as being larger than those of SILT, yet smaller than those of MUD.
See also SAND.

clipper Sleek, fast sailing vessel developed in the United States in the mid-1800s. To be considered a clipper ship, named for the way the ship "clipped off" the miles, the vessel had to possess a narrow hull, deeper in the stern than the bow, and have many sails rigged on at least three masts. From the launch of the first true clipper, the *Rainbow,* in 1845, to well into the 20th century, clippers were the "monarchs of the sea," until advancing technology introduced the more efficient steamship.

Cnidaria Animal phylum. *See* COELENTERATA.

cnidarian *See* COELENTERATES.

cnidocytes Stinging cells. *See* COELENTERATES.

CNODC *(China National Oceanographic Data Center)* A society responsible for organizing studies on China's marine policies and projects, collecting marine data, and exchanging findings with other countries.

coast The end of land and the beginning of the sea, or the other way around if one prefers. It is defined as being a border between land and water, often littered with debris washed in from the sea.

coastal plain A flat, landward extension of a submerged continental shelf bordering the sea made up of deposited sediments from rivers and seafloors, exposed over the vast generations by a rise in land level or a fall in sea level.

coastal zone management Innovative CZMs, (coastal zone management programs) are designed to protect coastal resources and ensure wise development along the coast. The first state CZM program began in California in 1972.

See also Useful websites in the Appendixes.

coastline Synonym of COAST.

CoastWatch A program of NOAA (the National Oceanic and Atmospheric Administration), established to focus on unusual environmental events, such as RED TIDES, and to map changes in tidal wetlands. The CoastWatch Program makes satellite data products and in situ data available to federal, state, and local marine scientists and coastal resource managers.

See also Useful websites in the Appendixes.

coccolithophorids Unicellular, eukaryotic members of the phytoplankton that have calcareous, buttonlike structures, or coccoliths.

coccoliths Chalky, calcified scales or disks protecting microscopic planktonic algal organisms called coccolithophorids. Coccoliths compose a large part of the ocean ooze and are an important component in chalky, sedimentary rocks. England's White Cliffs of Dover are a perfect example of fossilized coccoliths.

cockle Marine bivalve related to the clam, possessing a heart-shaped shell with riblike ridges radiating upward from the hinge. It uses its foot to burrow and propel itself through and over the sand. Cockles make up the genus *Cardium* in the family Cardiidae, order Eulamellibranchia, class Pelecypoda, phylum Mollusca, kingdom Animalia.

coelacanths A group of lobed-fin fossil fishes that first appeared around 350 million years ago and were thought to be extinct until one was caught in a fishing net in the early 20th century. Only one example of this species of bony fish exists today. Or does it? Coelacanths belong to the family Latimeriidae, subclass Crossopterygii, class Osteichthyes, subphylum Vertebrata, phylum Chordata, kingdom Animalia.

Coelenterata *(Cnidaria)* Phylum of the hydrozoans, siphonophores, hydrocorals, true jellyfish, box jellyfish, sea anemones, and corals. These fall within the kingdom Animalia.

coelenterates Also called cnidarians, these are marine animals of which there are two types, polyps and medusa. Examples are jellyfish, corals, Portuguese man-of-war, hydra, and sea anemone. All coelenterates are of symmetrical body formation, more or less, and do not have a separate circulatory system, an anus, or a gastrovascular cavity. Those that have mouths excrete back through them. Many possess stinging cells in their tentacles called cnidocytes, which eject venom into captured prey. Coelenterates belong in the phylum Cnidaria (or Coelenterata), kingdom Animalia.

coelom Fluid-filled body cavity found in annelids (worms) separating the body wall muscles from the gut, which provides a sort of structural support so the body may flex.

cold light Term given to the light produced by BIOLUMINESCENT animals.

colonial organisms Animals or plants that flourish together as a group and, in some species, may be joined together. ALGAE, mosses, and BRYOZOANS are examples of colonial organisms.

comb jellies Marine invertebrates characterized as having a gelatinous body, radial symmetry, and eight rows of ciliary combs. *See* CTENOPHORE.

commensalism A type of nonharmful, food-sharing SYMBIOSIS between two nonparasitic animals of different species. Animals that practice commensalism are called commensals. Some are free to come and go independently of each other. Others may never separate, such as the coral polyp and its zooxanthellae.

See also REMORA.

community Term used to describe plant and animal populations that relate to each other ecologically in the same region, as opposed to an assemblage, which are plant and animal populations that do not relate to each other ecologically in the same region.

compensation depth Also known as compensation light intensity, that point in the WATER COLUMN where the amount of light produced by PHOTOSYNTHESIS exactly balances respiratory losses in plants. In other words, the oxygen demand equals the oxygen generated. Below the compensation depth, plants are not able to survive.

complex camera eye The type of image-forming eye found in squids and octopuses, with an iris diaphragm and variable focusing.

compound eye An eye consisting of many individual lens systems (ommatidia), found in many ARTHROPODS.

concentric Sharing a center. In BIVALVES, it refers to lines or ridges that form an arching pattern.

conchology A branch of zoology that deals with the study of MOLLUSK shells.

conic projection Map projection of a globe placed on paper as a cone. Though its design shows distances accurately, it is not widely used in mapping because of the relatively small zone it represents. In a conic projection, parallels are projected as concentric arcs of circles, and meridians as straight lines radiating from the apex of the flattened cone.

conservation The protection of our global environment through the careful management of Earth's resources.

conspecific Different form in the same species, for example, the CUTTLEFISH versus the OCTOPUS.

consumer Animal that feeds on plants (primary consumer) or on other animals (secondary consumer).

continent A major landmass of Earth, as opposed to an ISLAND, which is a minor landmass of Earth. Land constitutes around 29 percent of the Earth's surface.

continental drift Name of the theory introduced in 1912 by ALFRED WEGENER, stating that the continents have traveled (and are still traveling) over great distances throughout the life of Earth and in the beginning belonged to one great supercontinent, which he named PANGAEA. Though at first disputed, as are so many other theories of new discovery, the theory is supported through evidence provided by the discovery of PLATE TECTONICS.

continental island *See* ISLAND.

continental margin The undersea edges of continents, consisting of the continental shelf, continental slope, and continental rise.

continental rim The rim of the continental shelf.

continental rise The gentle rise, deep at the base of the continental slope, consisting of SEDIMENT that has slid down from the slope.

continental sea A shallow sea submerging a low-lying part of a continent.

continental shelf A shallow offshore seabed that extends from the low watermark of a shoreline to the shelf break of the upper continental slope.

continental slope A steep slope beginning at the shelf break, or rim, at around 500 to 700 feet (152–213 m) that extends down to the continental rise or an ABYSSAL PLAIN.

convergence The region where different water masses come together, often resulting in the sinking of surface water.

Cook Inlet A northwestern arm of the Gulf of Alaska centered at about 60° N and 152° W. It separates Alaska's Kenai Peninsula from its southern continental coast. Cook Inlet has the second-highest tidal range in the world; the largest is that of the Bay of FUNDY.

Cook Strait Named for its founder, the English navigator CAPTAIN JAMES COOK, this is a passage of water separating the North and South Islands of New Zealand between the TASMAN SEA and the Pacific Ocean. It is centered at about 41° S and 175° E.

copepod Small, shrimplike member of the zooplankton.

coral Group of benthic marine invertebrate cnidarians (coelenterates) characterized by their protective horny skeletons consisting of secreted CALCIUM CARBONATE. Often they exist in large colonies. Many feed at night by extending soft tentacles from their hard skeletons and seizing animal plankton that washes up against them. Coral belong to the class Anthozoa under the phylum Cnidaria, kingdom Animalia.

coral (Courtesy of NOAA)

coral bleaching A phenomenon whereby coral reefs expulse zooxanthellae and appear to bleach as a consequence of environmental stresses, such as pollution or abnormally high water temperatures such as those resulting from an EL NIÑO year.

coral head A large section of coral that grows to the surface of the sea.

coral island *See* ISLAND.

coral reef A shallow, submerged organic platform of the Tropics composed of nothing but coral, coral sand, and their symbiotic algae, built up over generations by the living organisms growing on the skeletons of the dead. Coral reefs provide important ecosystems for the survival of other marine life.
See also ATOLL; BARRIER REEF; FRINGING REEF.

coral rubble Just what you might expect: coral fragments.

Coral Sea A sea belonging to the South Pacific Ocean centered at about 20° S and 155° E. The TASMAN SEA borders it to the south, New Guinea to the north, Australia to the west, and the Pacific Ocean to the east.

Coral Sea Basin An oceanic deep of the South Pacific's CORAL SEA centered at about 15° S and 152° E off the northeast coast of Australia. Its greatest depth is roughly 14,700 feet (4,480 m).

cord grass Salt-tolerant grass, species of the genus *Zostera,* inhabiting salt marshes.

CORE *(Consortium for Oceanographic Research and Education)* A Washington, D.C.–based association of United States oceanographic research institutions, universities, laboratories, and aquariums.
See also Useful websites in the Appendixes.

Coriolis effect In relation to marine science, the tendency for global currents to be deflected westward by the eastward rotation of the Earth. Wind-driven currents in the Southern Hemisphere are deflected to the left of the wind direction, or southwest, and wind-driven currents in the Northern Hemisphere are deflected to the right of the wind direction, or northwest.
See also CORIOLIS, GASPARD GUSTAVE.

cosmoid scales

cosmoid scales Similar to armor plates, they are bony, prehistoric-style SCALES with a smooth, rhomboid (uneven sides) shape, found on lungfishes and COELACANTHS. They are likely to have evolved from the fusion of PLACOID scales.

cosmopolitan species A species of plant or animal that is globally distributed.

cotidal chart A chart of a given area showing all points at which high tides occur simultaneously.

countercurrent An ocean current that flows directly back into another current.

counterillumination The emission of light by a midwater organism to match the light that surrounds it.

course The heading, or direction, in which a ship is steered.

crab Crustacean characterized by having a broad, flat body and five pairs of walking legs. Most species are benthic marine, but there are some that are freshwater and still others that venture onto land. There are around 4,500 species of "true" crabs and about 1,400 species of their cohorts, the HERMIT CRABS. Their prey choice is diverse, from detritus to plankton to live prey. Most crabs are edible: rich in protein and low in fat. Along with its cousin the king crab, the Bairdi crab is a commercially important species and is harvested yearly from the BERING SEA region. Crabs belong to the order Decapoda, class Malcostraca, subphylum Crustacea, phylum Arthropoda, kingdom Animalia.

cranium The major part of the skull, not including the jawbone.

crest The highest point of a wave.

Crete, Sea of Southern part of the AEGEAN SEA centered at about 36° N and 25° E. The Cyclades islands border it to the north, Crete to the south, the IONIAN SEA to the west, and Rhodes to the east.

Cromwell Current First recognized in 1952, it is an eastward-flowing undercurrent of a westward-flowing surface current of the equatorial Pacific Ocean.

Crozet Basin A basin of the Indian Ocean centered at about 40° S and 65° E, just north of an invisible line defined by the Prince Edward, Crozet, and Kerguélen Islands. It is bounded east and west by the Southeast and Southwest Indian Ridges, and the point of greatest depth is roughly 15,060 feet (4,590 m).

crustacean Successful group of arthropod marine invertebrates that include the CRAB, LOBSTER, and SHRIMP. All are decapods, meaning they each have five pairs of thoracic legs and a fused cephalothorax protected by a chitinous carapace. The small varieties strain food particles from the water (filter feed); the larger may be scavengers, omnivores, or predators; and it can be said of these animals that they are as abundant to the sea as insects are to land. Crustaceans belong to the subphylum Crustacea, phylum Arthropoda, kingdom Animalia. There are eight subclasses of Crustacea and 25,000 species.

Cryptophyta Phylum of the biflagellates and phytoflagellates, within the kingdom Protozoa.

ctenoid scales Rough, saw-edged SCALES with points on the surface. They are found on fish such as bass and perch.

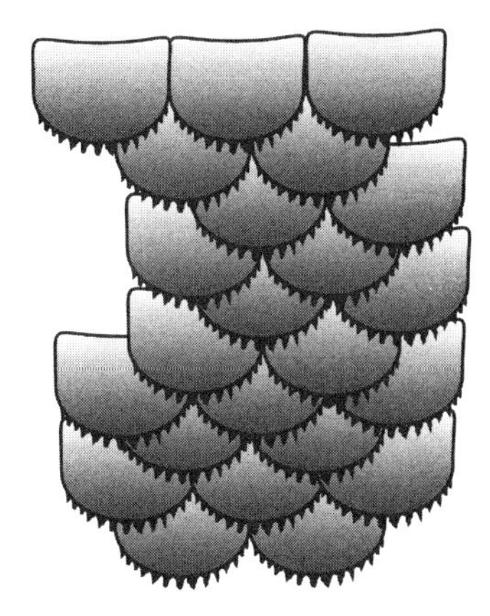

ctenoid scales

Ctenophora Phylum of the comb jellies, within the kingdom Animalia.

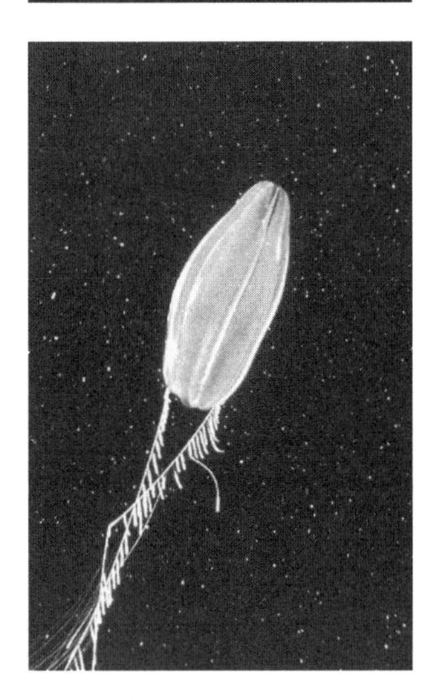
ctenophore (Courtesy of NOAA/NURP)

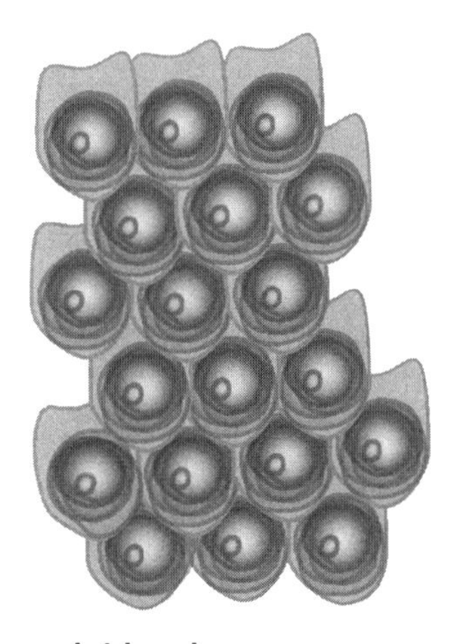
cycloid scales

ctenophore Transparent, planktonic animal, spherical or cylindrical in shape, with rows of cilia. The comb jellies are ctenophores, and most species have two long tentacles at each side of their body, which are used to capture prey. Ctenophores make up the phylum Ctenophora, formerly regarded as a class of the phylum Coelenterata.

current The movement of water across the surface of the world, which can occur at any depth, affected by wind, landward topographic features, rotation of the Earth, and water temperature.

cutter A small, lightweight vessel built for speed. There are several types of cutter; all are characterized as fast, comparatively small ships having a deep, narrow hull, and if sailing vessels, a long bowsprit and single mast rigged fore and aft as a sloop. The United States Coast Guard employs many diesel-powered cutters in its fleet.

cuttlefish A marine mollusk invertebrate of the squid class that lives deep in the sea but is sometimes found near shore. Cuttlefish are remarkable for their intelligence compared to that of their cousins. They have an internal shell called a cuttlebone (in squids it is called a pen), which is spongy and chalky; eight short arms; and two longer tentacles they use to capture prey. Cuttlefish belong to the family Sepiidae, suborder Decapoda, order Dibranchia, class Cephalopoda, phylum Mollusca, kingdom Animalia.

Cyanophyta Phylum of the blue-green algae, within the kingdom Plantae.

cycle A repeated sequence of events. *See* CARBON CYCLE; HYDROLOGICAL CYCLE.

cycloid scales Smooth, rounded SCALES of more ancient fish species, such as the salmon and whitefish.

cyclothem The cyclical pattern of marine (or nonmarine) coal-bearing strata deposited by the advance and retreat of the sea during a single cycle of sedimentation.

cytology The scientific study of cell structure in plants, animals, or nonliving matter.

dactylozooid A protective or defensive POLYP within a colonial cnidarian (COELENTERATE).

Davidson Current A deep countercurrent of the North Pacific Ocean setting northward between the CALIFORNIA CURRENT and the coasts of California, Oregon, and Washington during winter.

Davis Strait A body of water between BAFFIN BAY and the LABRADOR SEA, all belonging to the North Atlantic Ocean. It is centered at about 57° W along the imaginary Arctic Circle, separating Baffin Island from Greenland. Davis Strait is named for the English explorer JOHN DAVIS, who sailed through it in 1587.

davit On a ship, the mechanical arm that extends over the side or stern of a vessel that lowers or raises a smaller boat. Davits are also found on docks and piers.

davit (Photograph by Bob Williams/Courtesy of NOAA)

Dead Sea An isolated salt lake located at 32° N and 35.5° E along the border between Israel and Jordan. Because of its rapid rate of evaporation, its water composition is among the densest on Earth. Biblical names for the Dead Sea include Salt Sea, East Sea, and Sea of the Plain.

dead reckoning A method of reckoning (determining a ship's position) used when navigating on large, open bodies of water, not by using astronomical observations, but by observing a vessel's courses and distances by the ship's log. The advancement of the ship's position on the chart is determined by keeping careful record of the last accurate location using the course(s) steered and speed through the water. No allowance is made for the possible effects of wind, waves, current, or steering errors.

dead water phenomenon V. W. EKMAN was, in 1904, the first to perform systematic studies of this phenomenon. It occurs when a thin layer of freshwater (of lighter density) spreads over the surface of a salty sea (of heavier density), and then beneath the apparently calm surface the boundary between the two roils with great turbulence. It is exceptionally hard to propel a ship of size, by whatever means, through this region of differing density layers because of the drag caused by the transfer of the ship's own momentum to the waves.

decomposer Any organism that breaks down dead organic matter into inorganic components.

See also DETRITIVORE.

Decapoda The best known order of the Crustacea, which includes the edible species of the CRAB, LOBSTER, crayfish, PRAWN, and SHRIMP. Perhaps one day as a restaurant owner you might offer the "Decapoda Plate" on your menu, purely to amuse your clients!

decapods *See* CRUSTACEAN.

decompression sickness *See* BENDS.

deep convection The sinking of surface waters to form deep water masses, of which there are two types: a) convection (circulation due to outside force) near an unrestricted boundary where the dense water mass formation reaches the ocean bottom by descending a continental slope and b) convection in the open ocean, where the sinking occurs far from land and is predominantly downward. This is an important process in the maintenance of the ocean climate.

deep-sea swallower *See* BLACK SWALLOWER.

deep scattering layer Found in most ocean waters is a layer of diurnal (daytime) organisms—such as schools of small fish, squid, and larger crustaceans—that scatters sound waves as a result of gas bubbles within the animals. The layers range anywhere from 600 to 2,400 feet (182–731 m) deep, are between 150 and 600 feet (46–183 m) thick, and can range horizontally for many miles.

deep-water wave An ocean wave that exists at a depth greater than one-half its wavelength.

Delaware Bay A bay of the eastern North American continent belonging to the North Atlantic Ocean centered at about 39° N and 75° W between New Jersey and Delaware.

delta A wedge- or fan-shaped deposit of sediment at the mouth of a river that is built out into a lake or sea.

Demerara Abyssal Plain An oceanic plain situated between the Nares Plain and CEARA ABYSSAL PLAINS, which belong to the Atlantic Ocean's Guyana Basin. It is centered at about 8° N and 48° W between South America's northeastern continental shelf and the Mid-Atlantic Ridge.

demersal Living on or near the seafloor.

demineralize To remove salts and other minerals from water to create a relatively pure (distilled) product.

density 1) In a physical context the mass per unit volume. 2) In a biological context the number of organisms in a given space.

density current Flow of a heavier fluid beneath a lighter fluid, resulting from such causes as differences in water salinity, temperature, or content of suspended matter.

denticles *See* PLACOID SCALES.

deposit feeder An organism that feeds on organic particles it finds in seafloor sediment.

deposition Typically the result of flowing rivers—and to a somewhat lesser degree, weathering—this is the depositing of materials such as dust, sand, or gravel to form an accumulation of matter.

desalination Removal of salts from water.

desiccation The drying out or fatal water loss of an organism.

detrital The quantitative component of three major deep-sea sediments; the others are AUTHIGENIC and BIOGENIC. It derives from land materials, mostly in the form of aluminosilicate minerals, and is deposited in the ocean by rivers and wind.

detritivore Animal that feeds on detritus, such as the gastropod and prawn. *See also* DECOMPOSER.

detritus General term for fragments or grains worn away from rock by means such as abrasion or crumbling.

dhow Traditional, shell-built Arabian sailing vessel, the earliest of which were simple dugouts with wooden (usually teak) planks sewn to their sides to form a hull. The dhow is shallow, double-ended, one- or sometimes two-masted, and rigged with lateens (triangular sails). The Arabians use them for pearling and fishing.

diagenesis The range of changes sediment undergoes after its initial deposit, which includes compaction, cementation, authigenesis, replacement, crystallization, leaching, hydration, and bacterial action.

diatom A unicellular alga, of which there are about 8,000 species, which is a major component of plankton on which marine life depends. Diatoms—known formally as Bacillariophyceae—enjoy water low in salinity, usually reproduce by cell division, and are impregnated with silica, which constitutes a large part of the siliceous deposits known as diatom ooze or diatomaceous earth. Diatoms belong to the phylum Chrysophyta, kingdom Plantae.

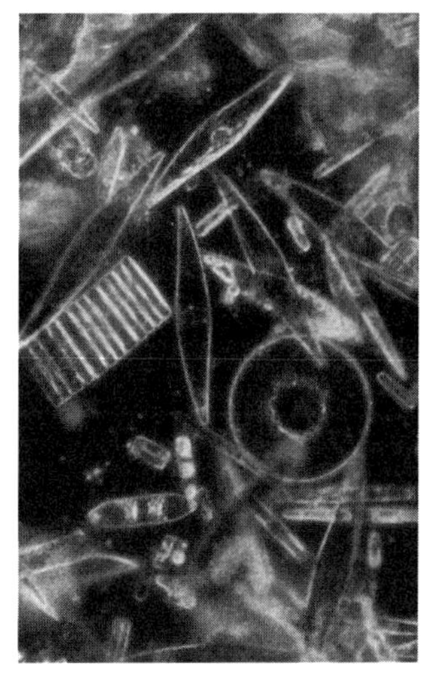

diatom (Photograph by Dr. Neil Sullivan/Courtesy of NOAA)

dicothermal layer A cold, vertical ocean layer of the high northern latitudes between the halocline and the thermocline running from 175 to 300 feet (53–91 m) in depth, allowing the water column to remain stable as a result of the salinity gradient's countering the imbalance effected by the temperature gradient.

diffraction The bending of a wave—it may be light or sound—around an object.

dioecious In the majority of sea cucumbers, having separate sexes and a single gonad.

dinghy A small open boat, often used as a lifeboat for a larger vessel.

dinoflagellates Unicellular marine organisms with two uneven flagella for locomotion that lend a distinctive spin to their movement. Commonly known as algae, they are microscopic protists, often autotrophic (photosynthetic), but can be heterotrophic (eating other organisms) as well. They form a significant part of primary planktonic production in the ocean. These organisms belong in the phylum Dinoflagellata (also called Pyrrhophyta in some classification systems), kingdom Plantae.

dispersion In oceanography, waves from the same location but originating with different frequencies spread out, or disperse, at different speeds across the water.

disphotic zone The ocean's "twilight zone," below the EUPHOTIC ZONE, where only small amounts of light can penetrate and PHOTOSYNTHESIS does not occur.
See also APHOTIC ZONE.

displacement The weight of water displaced by a floating object, and therefore, the object's weight.

diurnal 1) Event that occurs once a day. 2) Characteristic of animals of being active during sunlight hours.

diurnal tide A tide exhibiting one high and one low throughout the course of a single day.
See also SEMIDIURNAL TIDE; TIDES.

dive bell (Courtesy of NOAA/NURP)

dive bell A small, dome-shaped chamber useful for underwater work, open on the bottom and supplied with a limited amount of air under pressure, which keeps the water out. It is made from materials resistant to the effects of pressure under depth. Some have communication systems linking them to the surface. Certain types of diving bells can be used to transport commercial divers to underwater workstations. Others are operated from topside to supply a diver with unlimited air, but this system keeps the diver on a tether. Used in ancient times, the dive bell was the first invention to allow a human to breathe underwater.
See also EADS, JAMES.

divergence A horizontal flow of water (or air) away from a central region, often resulting in upwellings.

Dogger Bank An area of shallow sand in the center of the NORTH SEA at about 55° N and 2° E between Great Britain and Denmark. Average depth is about 55 to 120 feet (17–37 m), and it is famous not only for its profitable fishing but for the battle of Jutland, which was waged there by the British and German fleets during World War I.

doldrums Belts of air rising between the trade winds, usually occurring between 10° to 15° north and south of the equator. It can be a region of considerable calm and mild winds. Early sailing vessels would often be stranded in the doldrums with not a freshening of air to drive their ships for weeks on end.

dolphin 1) A nautical term describing a group of piles (stout poles) driven close together into the water and bound with cable or secured with steel and bolts to create a structure on which aids to navigation are typically mounted. 2) Marine mammal related to the whale and porpoise. The dolphin is a strong swimmer (nekton) that ranges in size from less than four feet (1.2 m)—characteristic of the tucuxi dolphin—to 10 feet (3 m)—characteristic of the bottle-nosed dolphin. Their diet consists mostly of fish and squid. Dolphins possess the distinctive "beak"-shaped mouth, a feature porpoises lack. ORCAS, or killer whales, are in the dolphin family, as well as the pilot whale. Dolphins belong to the suborder Odontoceti, order Cetacea, class Mammalia, phylum Chordata, kingdom Animalia.

dolphin (Photograph by Scott McCutcheon)

dorsal fin On a fish or marine mammal, the topmost fin on the animal's back. *Dorsal* means "upper surface" or "upper plane."

dory A small craft with a shallow draft and high, flared sides; it may be a motorboat, a rowboat, or a single- or double-masted boat rigged with lateens (triangular sails). Prevalent around the early 1900s, particularly on the eastern United States's Potomac River, this workboat was used in oystering, fishing, and hauling cargo. Dories can be dropped from the decks of larger fishing boats and used to set lines.

dory fish A food fish existing in both Atlantic and Pacific waters. Its characteristics are a deep-body compressed shape, large eyes, long dorsal spines, a large mouth with a protractile upper jaw for capturing smaller fish, and a dark spot behind each of the gills. Dories belong to the family Zeidae, in the order Zeiformes, infraclass Teleostei, subclass Actiopterygii, class Osteichthyes, phylum Chordata, kingdom Animalia.

double tide Also called double high water, agger, or gulder, this is a double-headed TIDE, that is, a high water consisting of two maxima of about the same height separated by a relatively small depression, or a low water consisting of two minima separated by a relatively small elevation.

Dover, Strait of A body of water separating England, to the west, from the continent of Europe, to the east, and connecting the English Channel, to the south, with the NORTH SEA, to the north. It is centered at about 51° N and 1.5° E.

downwelling The sinking of water below the surface, taking with it important oxygen to the lower depths. The TRADE WINDS, along with the CORIOLIS EFFECT, can cause downwellings along eastern coasts of continents.

See also UPWELLING.

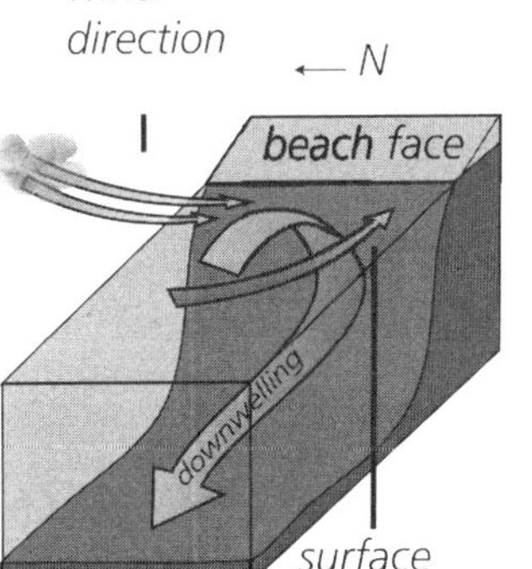

downwelling

draft The depth of water a boat in which a boat floats, measured from the surface of the water down to the ship's keel.

Drake Passage Also called Drake Strait, it is a narrow waterway centered at roughly 60° S and 60° W, separating the tip of South America from the South Shetland Islands of Antarctica and connecting the South Atlantic and South Pacific Oceans. It is named for its founder, SIR FRANCIS DRAKE, who, because of a storm, discovered it by accident in 1578.

dredge Floating device used for removing underwater sediment from canals, rivers, and harbors and usually held in place by supports that reach to the bottom of the body of water being worked. Dredging is important to navigation and sanitation, among other areas.

drift ice Floating ice floes or bergs borne over great distances by wind and current. *See* ICEBERG.

drogue Any device that is streamed astern to keep a vessel's speed in check or to keep its stern up to the waves in a following sea.
See also SEA ANCHOR.

drop-off A steep underwater cliff or precipice.

drowned valley A coastal upland inlet submerged, or drowned, after the sinking of land or the rising of sea level.

dry dock Term used when a ship is taken out of the water, usually for repairs. A temporary dry dock may consist of a series of "bents," or beams, that are submerged at high tide, allowing ships to pass over, and exposed at low tide, leaving ships high and dry and ready for quick inspection before the tide returns.

dugong *See* SIRENIAN.

earthquake Phenomenon that occurs when the Earth's crustal plates suddenly move together; as one plate slides beneath another, subducted (detracted, removed) into the Earth's molten interior; earthquakes and sometimes volcanoes result. In the ocean, TSUNAMI waves are sometimes generated as a result of water's poor ability to disperse shock waves.

East Arabian Current A strong, upwelling current setting northeastward along the Arabian coast. It materializes during monsoon season between the months of April and October.

East Auckland Current An extension of the East Australian Current circulating in a more or less geographically stable counterclockwise eddy east of New Zealand. During summer, part of it extends northerly into the East Cape Current and in winter it cycles into open ocean.

East Australian Current A western boundary current of the South Pacific Ocean setting west-southwest between the Great Barrier and

Chesterfield Reefs. It is the weakest of the world's boundary currents; however, it is prone to instabilities of stronger current and has been called the Gulf Stream of the Southern Hemisphere.

east boundary currents Surface currents that flow toward the equator on the eastern side of an ocean, such as the CANARY, CALIFORNIA, PERU, WEST AUSTRALIAN, and BENGUELA CURRENTS.

easterlies Winds blowing from the east, such as the equatorial and polar easterlies. Around latitude 30° in the Northern Hemisphere, steady northeast TRADE WINDS blow. Around latitude 30° in the Southern Hemisphere, southeast trade winds blow.
See also WESTERLIES.

East Greenland Current A current of the North Atlantic Ocean setting southward and then southwestward along the east coast of Greenland.

East Indian Current A seasonal current forming within the western end of the Bay of Bengal and setting northward between the months of January and October, countering the direction of the prevailing winds.

East Siberian Sea A shallow, marginal sea of the ARCTIC OCEAN centered at about 164° E and 73° N between the CHUKCHI and LAPTEV SEAS. Wrangel Island lies to the east, the New Siberian Islands to the west, Asia to the south, and the Arctic Ocean proper to the north. This sea is icebound most of the year.

East Wind Drift Westward flowing current close to continental Antarctica driven by polar easterlies.

ebb tide Movement of water away from the shore, resulting in low TIDE.

ecdysis In ARTHROPODS (and other animals), the periodic shedding of the exoskeleton.

echinoderm Slow-moving marine animal whose characteristics include a vascular system that utilizes seawater for respiration, locomotion, and reproduction. The echinoderm has no head, possesses radial body symmetry, and has arms in a pattern of five or multiples of five (pentaradial symmetry) equipped with "tube feet," which are suction devices that adhere to shells in order to pry them open. The word *echinoderm* is derived from the Greek words *echino* and *derm,* which when combined mean "spiny skin." Living examples are starfish, brittle stars, sea urchins, sand dollars, and sea cucumbers. Echinoderms make up the phylum Echinodermata in the kingdom Animalia.

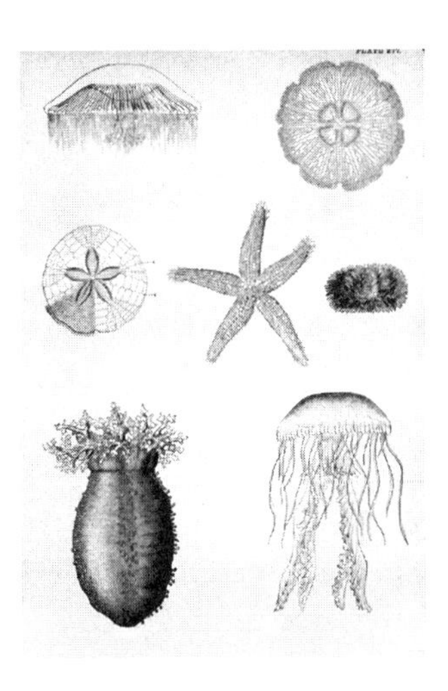

echinoderm (Courtesy of NOAA)

Echinodermata Phylum of the sea lilies, feather stars, starfish, brittle stars, sea urchins, sand dollars, heart urchins, and sea cucumbers. These fall within the kingdom Animalia.

Echiura Phylum of the spoon worms, within the kingdom Animalia.

echolocation Use of sound waves by some marine animals, such as orcas or bottle-nosed dolphins, to locate and identify underwater objects.

ecology Scientific study of living organisms and their effect on and interaction with each other and their environment.

ecosystem Term coined by the British ecologist Sir Arthur George Tansley in 1935 to describe the constant interchange of living organisms and nonliving parts of a particular environment, such as a CORAL REEF, desert, forest, or within the deep sea.

ectoderm In an embryo, the outer layer of cells engendering the newly forming organism's skin and nervous system.
See also ENDODERM; MESODERM.

eddy The spin-off flow of water from a rapidly moving current, resulting in small, gentle-looking whirlpools or large whirlpools that break up into smaller ones. In the ocean, eddies can be spun off by wind activity, upwellings, or a wave's encounter with an obstacle.

edge wave A wave traveling parallel to the coast with normal crests that diminish rapidly offshore.

eel Slippery, serpentine fish, usually scaleless, possessing dorsal and anal fins that run head to tail and average up to three feet (91 cm) in length. There are about 600 eel species, which can be of either the marine and/or freshwater variety, and many are edible delicacies. Types of eels are the conger, moray, American, European, deep-sea snipe, and deep-sea gulper. Eels make up the order Anguilliformes, infraclass Teleostei, subclass Actiopterygii, class Osteichthyes, phylum Chordata, kingdom Animalia.

eelgrass A flowering marine plant less commonly known as *Zostera,* which is a genus of plants that are neither grass nor seaweed. They have long, limp, ribbonlike leaves and can live for many years (perennials). Eelgrass is sometimes called a SAV or "submerged aquatic vegetation," plant and is important in providing a habitat for many marine species.

Ekman layer The thin surface layer of water riding horizontally across the ocean that is affected by wind.

Ekman spiral Named after the Swedish oceanographer VAGN WALFRID EKMAN, the gradual directional change over depth of ocean water due to the interaction of surface winds and the CORIOLIS EFFECT. A graph of the magnitude of the water movement and its direction versus the depth produces a spiral.

Ekman transport

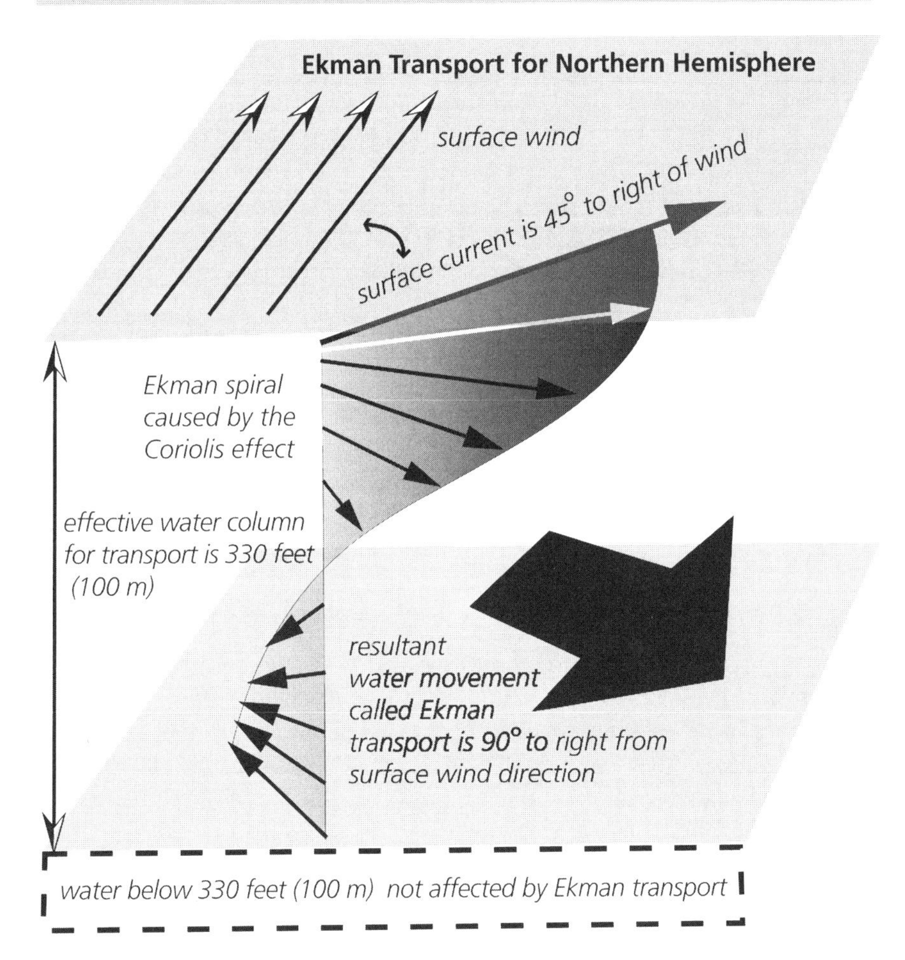

Ekman transport The result of the EKMAN SPIRAL, the sum transport of oceanic surface water as it moves with the wind, excluding the friction factor. The transport is at a 90° angle to the right (clockwise) of prevailing winds in the Northern Hemisphere and at a 90° angle to the left (counterclockwise) of the wind in the Southern Hemisphere.

electric fish Common name for a group of unrelated fishes capable of emitting an electric charge by way of specialized organs consisting of highly compact nerve endings. Electrical discharge is emitted to stun prey during a hunt, in self-defense, in the detection of prey or obstacles, and in navigation. Examples are the electric eel, electric catfish, and electric ray.

El Niño Sporadic warming of SST (sea surface temperature) across the Pacific Ocean caused by the weakening of equatorial easterly winds and precipitation. This warming slows the occurrence of nutrient-rich upwellings of colder deep water and leads to anomalies in global weather patterns.

Emperor Seamounts A chain of undersea mountains in the North Pacific Ocean trailing from the angled junction between the ALEUTIAN and KURIL TRENCHES off Siberia, all the way to the Hawaiian Islands.

endemic Organism found only in a specific area, region, or habitat and (reportedly) nowhere else.

Enderby Abyssal Plain Centered at about 60° S and 50° E, this is one of four oceanic plains encircling the continent of Antarctica; the others are the BELLINGSHAUSEN and WEDDELL ABYSSAL PLAINS and the SOUTH INDIAN BASIN. Depths average around 17,500 feet (5,334 m).

endoderm In an embryo, the deep, innermost, and primary layer of cells that develops for the newly forming organism the lining for its pharynx, respiratory tree, and digestive and excretory systems.
See also ECTODERM; MESODERM.

endolithic Living or growing between stones, as do the bryozoans.

English Channel An arm of the Atlantic Ocean centered at about 50° N and crossing the prime meridian. It separates France, to the south, from England, to the north, and connects with the NORTH SEA by way of the Strait of Dover to the east.

Entoprocta Phylum of marine sessile invertebrates, similar to bryozoans, within the kingdom Animalia.

environment All of the physical, chemical, and biological factors involved in the life of an organism or community.

epeiric sea Rarely encountered today, this is a shallow, broad inland sea; in Earth's past, (such as the early and middle Paleozoic), such seas were widespread. These seas were usually less than 800 feet (244 m) deep, with limited connection to the open ocean.

epibenthic 1) Pertains to the region of open ocean just above the ocean bottom, occupying about 10 feet (3 m) in vertical, physical space, and the creatures that live there. 2) Living just above the seafloor.

epicenter The point on the Earth's surface directly above an earthquake's point of origin.

epicontinental sea 1) Also known as a shelf sea, a shallow sea on a wide section of a continental shelf. 2) A shallow sea in the interior of a continent.

epifauna Descriptive of animals living attached to or moving freely over the sea bottom.

epilithic Descriptive of life existing on the surface of rock or other hard inorganic substrata.

epipelagic zone The open ocean zone closest to the surface where zooplankton and phytoplankton are the most abundant. The penetration of daytime sunlight determines the extent of an epipelagic zone; depending on geographic location, it can be anywhere from 300 to 700 feet (91–213 m). Examples of other living creatures found there are tuna, dolphins, manta ray, sailfish, and lanternfish.

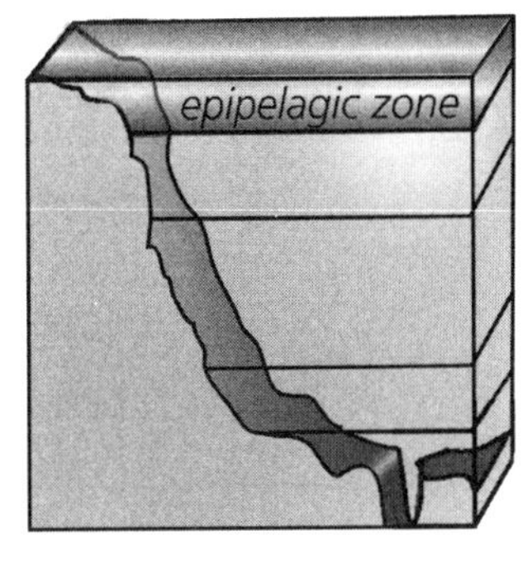

epipelagic zone

epiphytic Descriptive of a nonparasitic plant growing on the surface of another living plant.

epizoic Descriptive of a nonparasitic animal growing or living on the exterior of a usually larger animal.

equator Latitude 0°, an imaginary line around the center of the Earth, halfway between the North and South Poles, that divides the planet equally into the Northern and Southern Hemispheres.

Equatorial Currents Warm, wind-driven surface currents of equatorial Earth running westward. North and South Equatorial Currents are separated by Equatorial Countercurrents, which flow eastward.

equatorial zone The hottest of eight climactic zones of the Earth, lying between latitudes 10° north and south of the equator; the other zones are the: TROPICAL ZONE, Semiarid Tropical Zone, Arid Zone, Dry Mediterranean Zone, Mediterranean Zone, TEMPERATE ZONE, and POLAR ZONE.

equinox A term describing two days a year when the 12 daylight hours equal exactly the 12 nighttime hours at all points on the Earth. This occurs when the Earth's axis (center) is at right angles to an imaginary line between the Earth and the Sun. The equinox that occurs around March 21 marks the beginning of spring in the Northern Hemisphere and the beginning of fall in the Southern Hemisphere. Conversely, the equinox that occurs around September 22 marks the beginning of fall in the Northern Hemisphere and the beginning of spring in the Southern Hemisphere.

estuary A broad, low, semienclosed river mouth connected to the open sea where freshwater mixes freely with seawater. This term can apply to bays, gulfs, sounds, and inlets into which rivers empty and dilute salinity. In 1818 the English naturalist John Fleming detailed the mechanism of the mixing of waters in a well-mixed estuary, a study prompted by an early assumption that an "invisible layer" separated freshwater from salt water.

eukaryotic Pertains to organisms having cells with a distinct nucleus, such as PROTISTS, FUNGI, plants, and animals.

euphausiid Planktonic, shrimplike crustacean.

euphotic zone The ocean surface region where generous amounts of filtered sunlight allow the production of PHOTOSYNTHESIS.
See also APHOTIC ZONE; DISPHOTIC ZONE.

euryhaline An organism tolerant of wide ranges of salinity.

eurythermal An organism tolerant of a wide range of temperatures.

eustatic Pertains to the global change in sea level, caused most recently by the fluctuation of continental ice sheets during the last glaciations.

eutrophic Descriptive of waters containing a high concentration of nutrients. In some eutrophic situations, algal overgrowth can kill aquatic life and, in certain cases, poison drinking water by the release of plant toxins. This occurs when an increase of nutrients, such as nitrate and phosphate, is made available to the plants through the use of fertilizers or the burning of fossil fuels. This stimulated plant growth eventually causes local oxygen to be depleted, suffocating other life, and causes carbon dioxide levels to rise.
See also MESOTROPHIC; OLIGOTROPHIC; RED TIDE.

evaporation Changing of liquid into gas through heat, from the Sun or similar laboratory methods. When evaporation from open bodies of water occurs, the amount of vapor in the air increases until eventually it condenses and falls back to Earth in some form of precipitation.

evaporite A soluble, crystallized mineral (such as anhydrite, halite, or gypsum) precipitated from a quantity of liquid with a high content of that mineral.

evolution A gradual process of change, whereby groups of living organisms—or anything formed by atoms, even stars and planets—develop into a different, often better, species, sometimes as the result of an adaptation to environmental change.

excretion The elimination of waste products by animals, in the form of ammonia or urea and feces.

exoskeleton An animal skeleton on the outside of the body, protecting the soft tissue of the inside. Examples of animals with exoskeletons are CORALS, shellfish, ARTHROPODS, and insects.

extinction The total dying out of a species, caused by either natural selection or uncontrolled harvesting.

eye of the wind A nautical term meaning the direction from which the wind is blowing.

Falklands Current A cold, polar current of the South Atlantic Ocean setting northeastward along the east coast of Argentina. It meets with the BRAZIL CURRENT at about latitude 35° W and 40° S.

family In TAXONOMY, a classification group consisting of closely related genera, below an order and above a genus.
See also KINGDOM.

fast As a nautical term, this is said of an object that is secured to another.

fathom A term describing water depth measured in six-foot (1.83-m) increments, derived from a Danish word meaning "outstretched (arms)," which is supposedly the span of a man's outstretched arms from middle finger to middle finger. Though a term from the past, it is still used on modern charts today.

fault A FRACTURE ZONE in the Earth's crust, of which there are different varieties. Undersea faults are common, and new ones are still being discovered. In a strike-slip fault one zone slides horizontally past the other; it is seismically less threatening than a thrust fault, in which one side actually ramps up over the other.
See also RING OF FIRE.

fauna Term used to describe animals, particularly those in a selected region.

feces In animals, the solid form of waste produced through the process of excretion.

fecundity A term used when referring to the level or rate of egg or offspring production. High fecundity among marine fishes, such as cod and herring, which are an important part of the food chain, is a successful reproduction strategy.

felucca Small sailing vessel of ancient design. It has a narrow beam and shallow draft, is usually double-masted, and is rigged with lateen (triangular) sails. Once common in the eastern MEDITERRANEAN SEA, it is now usually found only on the Nile River.

femtoplankton The smallest of the marine planktonic organisms, 0.02–0.2 μm, in the family of marine viruses.

fetch The determination of wave height by the duration and velocity of wind over a length of water.

filament Chain of living cells.

filter feeders Marine animals that feed by sifting small organic particles from the water. Filter feeding is highly efficient and can allow a species to grow to its ultimate potential. Examples are CRUSTACEANS and BALEEN whales. Sponges and sea cucumbers use their entire bodies as a filter.
See also SIPHON.

fiord Also spelled *fjord,* a Norwegian word used to describe long, narrow, deep arms or inlets of the sea cut by rivers and glaciations. Most fiords have steep rock canyon walls, occasionally with spectacular waterfalls.

Finland, Gulf of A body of water centered at about 60° N and 25° E that extends eastward from the BALTIC SEA. It separates Finland to the north from Estonia to the south, with Russia spreading outward to the east.

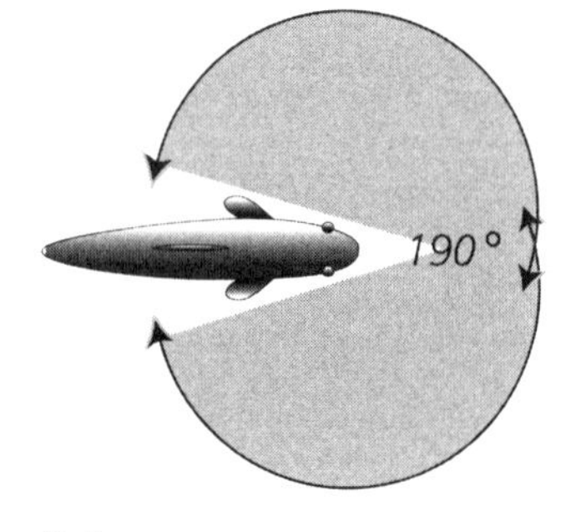

fish

fish Defined in the broadest sense, vertebrate, cold-blooded animals that live in water. Scientists divide fish into two basic groups, jawed and jawless, yet there are more fish varieties than there are land and water vertebrates combined. The smallest fish known is the pygmy goby of the Philippines, which grows less than 0.5 inch (1.3 cm) long. The largest fish is the whale shark, which is capable of growing up to 50 feet (15 m) long. Fish have no necks and cannot turn their heads; however, the placement of their eyes on each side of their body allows them a wide field of vision.

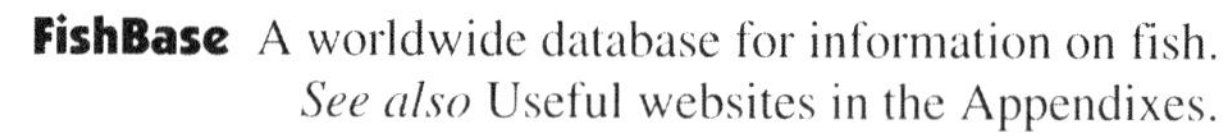

FishBase A worldwide database for information on fish.
See also Useful websites in the Appendixes.

flagellum Long, whiplike extension from a living cell's surface that provides the source of locomotion.

FLIP *(floating instrument platform)* A nonpropelled research barge owned by the U.S. Navy and operated by the Marine Physical Laboratory of SCRIPPS INSTITUTION OF OCEANOGRAPHY, acquired by them in 1964. It is towed to its site and then flipped vertical in the water, either to free-float or to hang moored while oceanographic research is being conducted. In this mostly submerged position it becomes a very stable platform and as such is capable of deploying an array of hydrophones for studying ambient noise and acoustic propagation in the ocean.

floe A relatively flat portion of free-floating ice. A floe can be anywhere from less than a foot to over a mile (30.5 cm–1,609 m) in diameter.

flood tide Movement of water toward the shore, resulting in a high TIDE.

flora Term used to describe plants, particularly those in a selected region.

FLIP (Floating Instrument Platform) (Photograph by Jurgen Gebhardt, *Stern* magazine)

Flores Sea A small sea amid the Indonesian Islands centered at about 8° S and 120° E. It lies among the JAVA SEA to the west, the BANDA SEA to the east, the Lesser Sunda Islands to the south, and the Celebes Islands to the north.

Florida Bay A bay in the Gulf of Mexico centered at about 25° N and 81° W between the tip of the Florida Peninsula and the Florida Keys.

Florida Current A warm-water current setting northward through the shallow, narrow Straits of Florida from the CARIBBEAN SEA, carrying a large water mass at good speed into the North Atlantic Ocean. This is a section of the GULF STREAM and extends from the Straits of Florida all the way to North Carolina's Cape Hatteras region.

flotsam Floating debris, especially that from a sunken ship. Compare to JETSAM.

flowmeter Instrument used for measuring the velocity of water. Some are placed on an ocean or river bottom or anchored at some depth above the bottom, and some are placed under BUOYS.

fluke 1) The tail of a cetacean, in the shape of a flattened heart. 2) A type of parasitic flatworm.

fluvial Pertains to processes related to flowing river water.

fluke (Photograph by Jan Straley)

fog A low-lying cloud that when over water makes navigation very difficult. Lighthouses are equipped with foghorns and bright, oscillating lights to aid seafarers in locating land. Ships in thick fog must comply with the navigation rules specific to their boating area and signal their position with a certain series of prolonged blasts from the ship's horn every few minutes; this signal advises other vessels of their location.

food chain Sequence of organisms in an ecological community in which each member is food for the next higher member in the sequence.

food web Complex of interrelated food chains within an ecological community; all the feeding relations of a community taken together; includes production, consumption, decomposition, and the flow of energy.

foraminifera A phylum of generally small-sized marine protists, having a jellylike body with delicate surface filaments capable of expansion and retraction and having calcareous chitinlike shells called tests, which are usually divided into perforated chambers. Their fossils form rich sedimentary deposits.

fore Nautical term, used to describe the forward area of a ship, or forward of amidships.

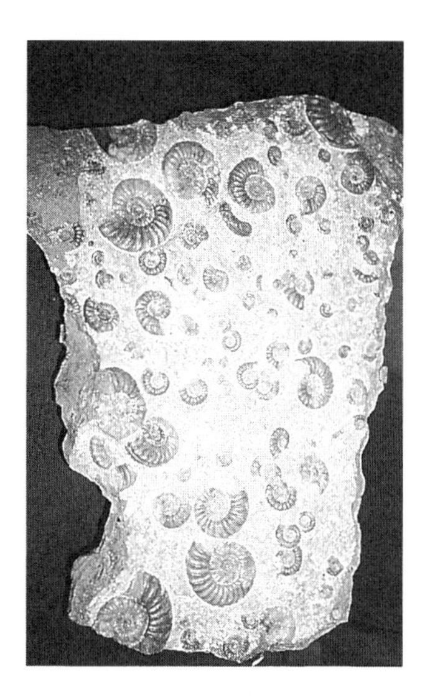

fossil (Photo courtesy: Royal Tyrrell Museum/Alberta Community Development)

foreflipper The front flipper of a PINNIPED.

foreland A section of land that juts into the sea.

foremast The forwardmost mast of a vessel.

foreshore zone Another name for the intertidal zone that lies between the low and high tide marks.

See also INTERLITTORAL *and* NEARSHORE ZONE.

forward As a nautical term, meaning toward the bow of a boat.

fossil The remains or trace of remains of an animal or plant from the ancient past, such as bones, shells, or leaf impressions preserved in rock. Shells, for example, can become preserved when the solution in which they rest evaporates, leaving a residue of dissolved mineral (usually CALCIUM CARBONATE or silica), which over time replaces their structure.

fracture zone A long, narrow break in the seafloor with a sheer drop-off over a mile deep (1,609 m). Numerous fractures cross the meandering path of the Mid-Atlantic Ridge and often offset the ridge's course. Fractures in the Pacific are sometimes fantastically long; they can extend for thousands of miles as they intersect the East Pacific Rise, offsetting its path as well.

Fram Basin The smallest of the four major basins of the Arctic seafloor beneath the ARCTIC OCEAN; the others are the CANADA, MARKOV, and

NANSEN BASINS. It is centered between the Markov and Nansen Basins and separates the Lomonosov Ridge from the Nansen Ridge. The greatest depth is roughly 14,230 feet (4,337 m).

freeboard On a ship's hull, the distance between the surface of the water and the gunwale.

freshwater Nonsaline water; water containing little or no salts. *See also* LIMNOLOGY.

frigate A style of navy ship that, as an early three-masted sailing vessel, carried a complement of war guns on its single deck and was more maneuverable than other ships of the line. In the War of 1812 between Britain and the United States, the British vessel *Leopard* fired on and boarded the American frigate *Chesapeake,* killing and wounding a number of American sailors. Today's frigate is a multipurpose warship, smaller than a destroyer or a cruiser and carrying less artillery.

fringing reef A CORAL REEF attached to and fringing a coast, with no lagoon separating it from the land. It is one of three distinct forms of coral reefs; the others are the ATOLL and the BARRIER REEF.

frustule The silaceous shell of a diatom.

fucoids Large brown seaweed groups belonging to the plant phylum Phaeophyta, kingdom Plantae.

fucoxanthin The characteristic red-brown pigment of brown algae that masks the plant's CHLOROPHYLL.

fully developed sea In the ocean, a hypothetical situation in which a state of maximal wave development is achieved as a result of storm duration and FETCH.

Fundy, Bay of A finger of the North Atlantic Ocean centered at about 45° N and 66° W. It separates Canada's Nova Scotia, to the east, from New Brunswick, to the northwest, and is open to the Gulf of Maine to the south. Within this body of water exists the highest tidal range in the world.

Fungi 1) Kingdom of either single-celled or multicelled organisms, the fungi and lichens, which include the phyla of Mycophyta and Lichenes. 2) The organisms themselves, which subsist on dead plant tissue and contain no CHLOROPHYLL.

Fury and Hecla Strait A shallow, barely navigable stretch of water centered at about 70° N and 84° W between Canada's Melville Peninsula and western Baffin Island. It is considered part of the NORTHWEST PASSAGE.

fusiform A shape, typical in fish, that is spindlelike or torpedolike, tapering at the rear and pointed in the front.

galleon A modified version of the CARRACK, redesigned by the English navigator John Hawkins, that eliminated the big bow structure and the large forecastle, which undesirably had always caught the wind. The modifications made the galleon more maneuverable than the carrack.

galley 1) Nautical term used for the kitchen aboard a ship. 2) Greek warship used before 3000 B.C.E. and retired in the 18th century. Older galleys had tiers of long oars along each beam (side) and a single mast rigged with a square sail. Its principal weapon was a ram of iron or bronze protruding from the bow.

gamete A mature reproductive male or female haploid cell (a cell containing a single chromosome of one type or the other) capable of initiating formation of a new individual after fusion with a gamete of the opposite sex.

ganglion Mass of nervous tissue containing many neurons and synapses, especially those that form outside the brain or spinal cord.

gangway On a ship, the area where people board and disembark. The temporary ramp on which people traverse between the vessel and the wharf is called the gangplank.

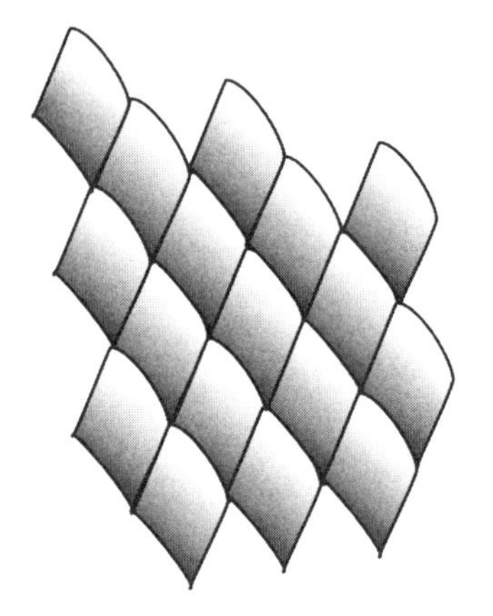

ganoid scales

ganoid scales Tough, diamond-shaped SCALES of less developed fishes, such as the sturgeon and gar.

gaping In BIVALVES, the description of valves that do not close completely.

Gaspé Current A current of the Gulf of Saint Lawrence, driven mainly by Canada's Saint Lawrence River. It flows past the Gaspé Peninsula, spreads fingers into the gulf, and then exits to the Atlantic Ocean through the Cabot Strait, where it merges with the Cabot Current.

gastropod The largest group of mollusks, which include snails, slugs, limpets, and abalones. These harmless, characteristically single-shelled animals have an astounding variety of colors. Second only to insects, they form the next largest class of the animal kingdom, are the only mollusks that can survive on land, and can be of the herbivore, carnivore, or omnivore variety. These creatures are in the class Gastropoda under the phylum Mollusca, kingdom Animalia.

Gastropoda The snail class, and tallying in at around 74,000 species, the largest class in the phylum Mollusca.

Gastrotricha Phylum of the microscopic flattened marine worms, within the kingdom Animalia.

GDC *(Global Drifter Center)* A Miami, Florida, data center that manages the deployment of drifting buoys around the world.
See also Useful websites in the Appendixes.

genus In TAXONOMY, a classification group consisting of closely related genera, below a family and above a species (plural, *genera*).
See also KINGDOM.

geochronology The scientific dating of events in Earth history by analyzing radioactive elements of corals and seawater.

geoduck Pronounced "goo '-ee-duck," and, no, it is not a sticky bird! A geoduck is a very large bivalve mollusk that can weigh sometimes more than 6.5 pounds (3 kg), with a shell measuring as much as seven inches (18 cm) long. These giant, deep-burrowing clams filter feed through the silt using long siphons that reach up to three feet (91 cm) in length. They can be found in the intertidal zones of Mexico all the way up to Alaska and are popular on the menus of seaside restaurants. Geoducks belong in the family Saxicavidae, order Eulamellibranchia, class Pelecypoda (or Bivalvia), phylum Mollusca, kingdom Animalia.

geographic north The direction toward the actual geographic North Pole from any random point on the Earth.
See also MAGNETIC DECLINATION; MAGNETIC NORTH.

geography Systematic scientific study of the surface features of the Earth.

Georges Bank A shallow area off the coast of New England centered at about 42° N and 68° W, once one of the world's most productive fisheries. Stocks today are now greatly depleted as a result of overfishing.

geology The study of Earth's rocks and minerals and their development.

GEOSECS *(Geochemical Ocean Sections Study)* A worldwide survey of the three-dimensional distribution of chemical, isotopic, and radiochemical tracers in the oceans. Atlantic expeditions took place from July 1972 to May 1973; Pacific, from August 1973 to June 1974; and Indian Ocean, from December 1977 to March 1978.

geostrophic force A force resulting from the Earth's rotation and the CORIOLIS EFFECT, used to account for a change in wind direction relative to the Earth's surface.

Gerard barrel A seawater sampling barrel made of stainless steel with the capacity to hold 66 gallons (250 l), which is the quantity required for making an accurate analysis.

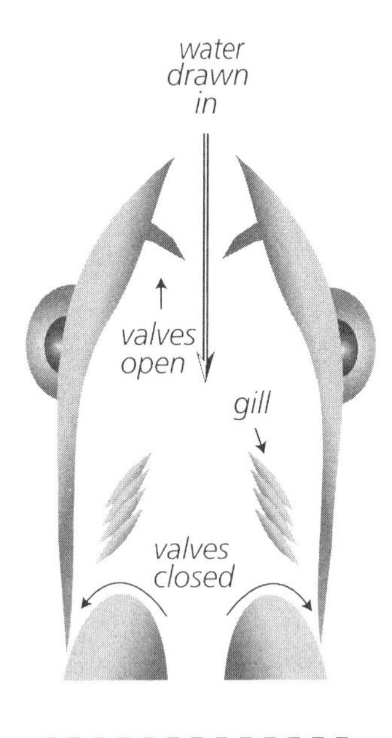

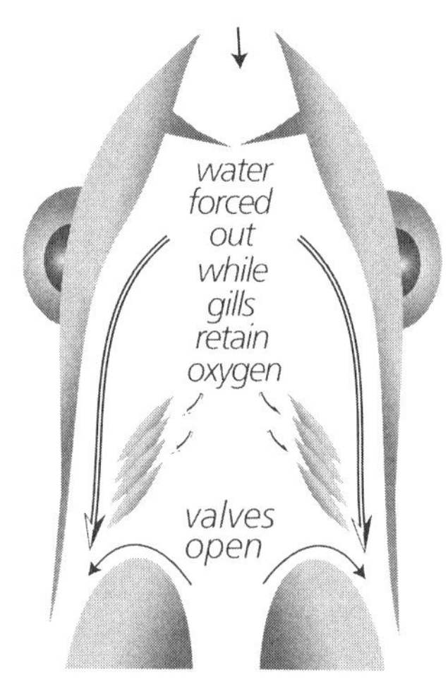

gill

GHCC *(Global Hydrology and Climate Center)* A research center whose objective is to address global hydrological processes.
See also Useful websites in the Appendixes.

giant squid Notorious marine invertebrate long believed to be mythical. Though it is not, so far it has never been seen alive. It skulks secretively at depths of 1,000 to 2,000 feet (305–610 m), where it falls prey to sperm whales. Sometimes squid rise too far to the surface and die. Because of this, specimens have been found as large as 60 feet (18 m) long. The Reverend Moses Harvey of Newfoundland was the first to put a giant squid on display after purchasing one from fishermen. Professor A. E. Verrill of Yale University then conducted the first scientific study of the huge creature. The giant squid is classified by the genus and species *Architeuthis princeps*, of the family Architeuthidae, class Cephalopoda, phylum Mollusca, kingdom Animalia.

Gibraltar, Straits of An ancient passage of water located at the western end of the MEDITERRANEAN SEA that allows circulation between the sea and the Atlantic Ocean. Centered at about 36° N and 6° W, it separates the southernmost tip of Spain from the northernmost tip of Morocco.

gill Respiratory organ in fish and other marine animals, such as mollusks, crustaceans, and some worms, used to obtain oxygen from the water. In fish, the gills are located immediately behind the eyes.

globigerina A unicellular PROTOZOAN living in surface waters of the sea.

globigerina ooze A calcareous ooze covering around 35 percent of the deep seafloor, found at depths around 16,400 feet (4,998 m) or more. This ocean ooze is composed of the shells of globigerina, creatures that enjoy the warmer waters of the Earth.

glycohalophytes Plants able to withstand moderate salinity.

glycophytes Plants intolerant of salinity.

GMT *See* GREENWICH MEAN TIME.

gnathostomata One of two superclasses of fish, those that have jaws (such as sharks, rays, and bony fishes). This is opposed to AGNATHA, fish without jaws (such as lampreys and hagfish).

Gnathostomulida Phylum of the small marine intestinal worms, within the kingdom Animalia.

Gondwanaland One of two supercontinental masses resulting from the breakup of the supercontinent PANGAEA; the other is Laurasia. From

it South America, Africa, India, Australia, and Antarctica were formed in Earth's Southern Hemisphere.
See also CONTINENTAL DRIFT; PLATE TECTONICS.

gorgonian corals Plantlike marine animals known as the horny CORALS, of which several species are known. These are flat, lacy, delicate corals of many different colors. Gorgonians thrive in tropical waters and are always found growing at right angles to the tides or currents, using their intricate skeletal network to catch planktonic nutrients from the water. Examples include sea whips, sea plumes, sea rods, sea fans, and precious corals. Gorgonians make up the order Gorgonacea, class Anthozoa, phylum Cnidaria, kingdom Animalia.
See also SOFT CORAL.

Gorlo Strait Centered at about 67° N and 42° E along the northern coast of Russia, this is a threshold of water between the Barents and White Seas of the Arctic Ocean.

GPS *(Global Positioning System)* A system of the United States government made up of 24 Navstar GPS satellites that provide highly accurate worldwide navigational and positioning information 24 hours a day. As the satellites constantly transmit their precise time and position in space, receivers on Earth may use a triangulation of the signals to pinpoint the receiver's geographic location.

Grand Banks Prime fishing grounds, this submerged, roughly triangular-shaped continental margin lies east of Nova Scotia, and southeast of Newfoundland and provides a major source of the North Atlantic's cod and herring catch. Average depth is 130 to 300 feet (40–91 m).

gravimeter A highly sensitive instrument used aboard research vessels to detect gravitational anomalies and, therefore, in conjunction with regional seismic readings, show the topographic feature of the seafloor.

Great Australian Bight A part of the Indian Ocean indenting the southern coast of Australia. It is centered at about 34° S and 130° E.

Great Barrier Reef A huge BARRIER REEF centered at about 18° S and 146° E existing off the northeast coast of Australia.

great circle The largest circle that can be drawn on a sphere, important to navigation since the shortest distance between two points on the Earth's surface is the great circle on which both points lie. The meridians of longitude and the equator are all great circles.

great white shark The most well known member of oceanic predators belonging to the mackerel SHARK family. One can safely say that for years this huge animal has been the ultimate purveyor of human fear within the ocean, considered to be nothing but a "killing machine." But thanks to advanced studies, science today has a better understanding of the great white, and it is slowly overcoming its reputation as a villainous hunter. In 1969 one of the largest specimens ever caught (up until that time) was taken off the coast of New York: 17.5 feet long (5.33 m) long and weighing 4,500 pounds (2,041 kg). Mackerel sharks belong in the family Isuridae, order Selachii, class Chondrichthyes, subphylum Vertebrata, phylum Chordata, kingdom Animalia.

See also CARTILAGINOUS FISH; PLACOID SCALES.

greenhouse effect The trapping of heat in the atmosphere by certain gases, which absorbs outgoing radiation and thus causes a rise in Earth's temperatures.

greenhouse gas Atmospheric gas of the Earth responsible for the greenhouse effect; such gases include carbon dioxide, methane, and CFCs (chlorofluorocarbons).

Greenland Basin A basin of the Arctic Ocean lying off the east coast of Greenland between the GREENLAND SEA and the NORWEGIAN SEA, home to the undersea Mohns Ridge. Within the basin are the Boreas and Greenland Plains, which are separated by the Greenland Fracture Zone. It is centered at about 75° N along the prime meridian and averages 11,000 feet (3,353 m) in depth.

Greenland-Scotland Ridge An undersea ridge in the North Atlantic Ocean running roughly east-west and centered at about 61° N and 11° E between the north coast of Scotland and the Faroe Islands.

Greenland Sea Centered at about 75° N and 10° W, this is one of the so-called NORDIC SEAS belonging to the ARCTIC OCEAN. The coast of Greenland borders it to the west, Iceland to the south, the NORWEGIAN SEA to the east, and Spitsbergen to the north.

Greenpeace A large international environmentalist group that formed almost by accident after a small band of people boarded a vessel in Alaskan waters to bear witness to nuclear weapon testing planned for Amchitka Island in 1971. Today the group's 250,000 U.S. and 2.5 million worldwide members work toward the preservation of Earth's environment by addressing issues such as preserving old-growth forests, halting global warming, exposing toxic pollutants, protecting the oceans, eliminating the threat of genetic engineering, and ending the nuclear age.

Greenwich Mean Time (GMT) Also known as Universal Time (UT), this is the mean solar time referenced along the zero line of longitude (PRIME MERIDIAN) that passes through Greenwich, England, and is used as the primary basis of standard time throughout the world. Navigators prefer the term *GMT;* astronomers prefer *UT.*

Greenwich Meridian *See* PRIME MERIDIAN.

grouper Large-jawed, heavy-bodied sedentary fish found inhabiting temperate and tropical coastal areas close to protective weeds and corals. Most can change their coloration. Groupers belong to the family Serranidae, order Perciformes, class Actinopterygii, subphylum Vertebrata, phylum Chordata, kingdom Animalia.

growth lines In shells, lines that mark the stages of growth.

groyne A beach fence erected to help prevent longshore drift from moving sand and shingle away from the beach.

Guiana Basin A basin of the mid-Atlantic Ocean centered at about 10° N and 50° W off the northern coast of South America. It includes the DEMERARA and CEARA ABYSSAL PLAINS and averages roughly 16,000 feet (4,877 m) at its deepest.

Guinea Basin A basin of the mid-Atlantic Ocean centered along the equator (0° latitude) and the prime meridian (0° longitude) off the central western coast of Africa. It includes the Guinea Abyssal Plain and averages roughly 17,000 feet (5,181 m) at its deepest point.

Guinea Current A warm, high-salinity, oxygen-poor current of the Atlantic Ocean that moves eastward along the bulge of the central western coast of Africa. It forms part of the GUINEA DOME and merges with the Equatorial Countercurrent.

Guinea Dome A consequence of cyclonic gyres in the mid-Atlantic Ocean off the western coast of Africa that forms during the summer months. The water along 10° N and 22° W domes as a result of seasonal upwelling of the thermocline.

gulder *See* DOUBLE TIDE.

gulf Coastal inlet larger than a bay that penetrates deep into land.

Gulf Stream A warm current of the North Atlantic Ocean first discovered by the Spanish explorer Ponce de León in 1513. The result of a northern, then western swing of the North Atlantic Current, it begins in the Gulf of Mexico, then exits as the FLORIDA CURRENT through the Straits of Florida and widens to approximately 50 miles (80 km) across along the southeast coast of the United States.

gulper eel Deep-sea marine EEL, also called a pelican eel, characterized by an enormous mouth, smooth, scaleless skin, and the lack of a swim bladder. Found in temperate and tropical BATHYPELAGIC waters, they are usually black, have a long whiplike tail, and can grow to 30 inches (76 cm) long. Gulpers make up the family Eurypharyngidae, order Saccopharyngiformes, class Actinopterygii, subphylum Vertebrata, phylum Chordata, kingdom Animalia.

gunwale The upper edge of a boat's sides.

Guyana Current A current of the mid-Atlantic Ocean setting northwestward along the eastern coast of South America at the equator. It is the flow of water between the CARIBBEAN and NORTH BRAZIL CURRENTS.

guyot Known also as a tablemount, this is a flat-topped undersea mountain first discovered by the American geologist HARRY HESS and named in honor of Arnold Guyot, a 19th-century geologist. Guyots are usually subsiding volcanic islands, leveled by the action of waves after they sink beneath them.

See also SEAMOUNT.

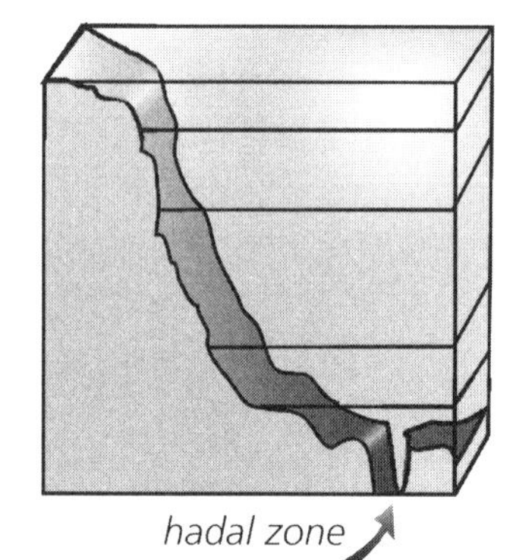

hadal zone

gymnosperm *See* ANGIOSPERM.

gyre In the open ocean, a roughly circular path of water often with a stagnant center. Gyres move clockwise in the Northern Hemisphere and counterclockwise in the Southern Hemisphere.

habitat Place in nature where plants and animals live and grow together, including the landscape and climate that affect them.

hadal zone A name applied to the deepest of all oceanic trenches where no light exists. These areas plunge from 20,000 to 36,200 feet (6,096–11,033 m) below the surface. Pressures are fantastically high, from 600 to 1,100 times greater than pressure at sea level. But these areas are rare, comprising barely over 1 percent of Earth's total ocean area. Animal life consists primarily of crustaceans, cnidarians (coelenterates), polychaetes, mollusks, and echinoderms. In 1960 the bathyscaphe *Trieste* made the first deep-sea dive into the world's deepest hadal zone, the Pacific Ocean's MARIANA TRENCH.

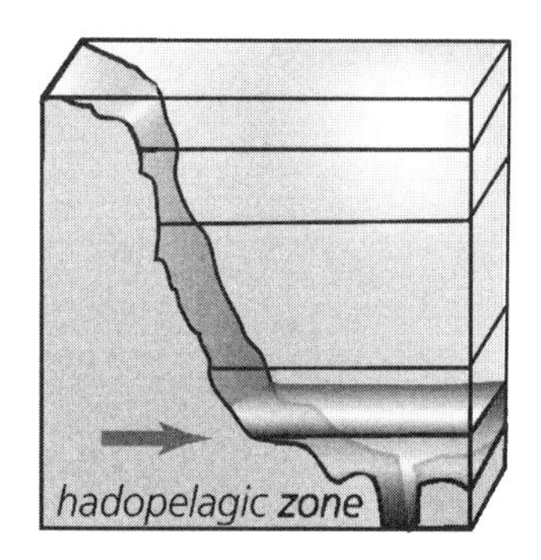

hadopelagic zone

hadopelagic zone Lightless area of open ocean beneath the ABYSSOPELAGIC ZONE that extends from around 20,000 feet (6,096 m) to the seafloor. Most of the life-forms are usually quite small, such as MEIOFAUNA, which live among the sediment grazing on bacteria, and deposit feeders such as polychaetes, bivalves, crustaceans, sea cucumbers, and brittlestars. In contrast, the warm-water oases surrounding deep-sea HYDROTHERMAL VENTS attract such life as gigantic worms, clams, and dense clusters of mussels, shrimps, crabs, and fishes.

GLOSSARY
gunwale – hadopelagic zone

Haida Current A seasonal, wind-driven surface current of the Pacific Ocean setting northward along the northwestern coast of North America between British Columbia and Alaska. It occurs during the winter months and is most predominant between November and February.

halibut A species of flatfish that is related to the flounder but is longer, thicker, and heavier. Flatfish are born as normal, bilaterally symmetrical fish, but as they mature, the left eye migrates to the right side of the body, their left side turns white and becomes their underside, and their right side remains a mottled brown-gray that blends well with the seafloor. Female halibut can grow as large as nine feet (2.7 m) long and weigh as much as 700 pounds (317 kg). The halibut is an excellent food fish, especially when fresh. Halibut belong to the genus *Hippoglossus*, family Pleuronectidae, order Pleuronectiformes, class Actinopterygii, subphylum Vertebrata, phylum Chordata, kingdom Animalia.

haline Salty, or related to SALINITY.

Halmahera Sea A small sea within the Indonesian islands centered along the equator at about 129° E. New Guinea borders it to the east, Halmahera to the west, the Ceram Sea to the south, and the PHILIPPINE SEA to the north.

halobates Known as ocean striders or water bugs, this genus of wingless bug is strictly marine and the only oceanic animal to live on the stratum of the sea surface. They never dive. Most are found in inland waters, but some are pelagic, laying eggs on anything that floats and preying on zooplankton, dead jellyfish, fish eggs, and larval fish trapped on the ocean surface. Halobates belong to the order Hemiptera, class Insecta, phylum Arthropoda, kingdom Animalia.

halocline Water in which the salt content changes rapidly with depth. The halocline forms the boundary between regions of differing salinity.

halophytes Plants that thrive in saline soils.

Haptophyta Phylum of the small phytoflagellates, including coccoliths, within the kingdom Protozoa.

Hatteras Abyssal Plain Centered at about 30° N and 74° W, this is part of the floor of the northwest Atlantic Basin, east of North America's continental shelf.

hawser A nautical term descriptive of a heavy rope or cable used for mooring or towing.

head As a nautical term, used to describe 1) a ship's toilet or 2) the upper edge of a triangular sail.

heading The direction in which the bow of a moving vessel is pointed at any given time.

headland A promontory or cape of land, typically with sheer cliff walls, that points into a sea.

headway The forward motion of a vessel; opposite of *sternway*.

Hemichordata Phylum of the acorn worms and pterobranchs, free-living or sessile tentacle-bearing animals, within the kingdom Animalia.

hemipelagic Pertaining to continental margins and the abyssal plains that border them.

hemisphere Applied to the sphere of the Earth, it is either the northern or southern half, as defined by the equator, or the eastern or western half, as defined by a meridian.

herbivore Animal that feeds only on plants.

hermaphrodite An organism possessing sexual characteristics of both male and female. Almost all flatworms are hermaphroditic.

hermaphrodite brig *See* BRIG.

hermit crab Family of armorless, mostly marine crustaceans that find protection by borrowing the abandoned shells of gastropod mollusks and then moving to larger ones as they grow. Marine hermit crabs make up the family Paguridae, order Decapoda, class Malcostraca, subphylum Crustacea, phylum Arthropoda, kingdom Animalia.

herring A common reference to several species of fish that sweep about the oceans in huge schools, whose individuals number in the billions. The herrings and smelts hold great economic importance to humans, being caught in billions each year and used for fertilizer, animal food, frozen bait for sport fishing, and oil. These fish, which feed on plankton, have high FECUNDITY, and are a dietary staple for other sea life. Herrings belong to the order Clupeiformes, class Actinopterygii, subphylum Vertebrata, phylum Chordata, kingdom Animalia.
See also MENHADEN.

heterotroph An organism unable to manufacture food from inorganic compounds and therefore requiring organic compounds for its food.

high tide *See* FLOOD TIDE; TIDES.

hindflipper The rear flipper of a PINNIPED.

hinge In BIVALVES, the connection that holds the two halves together and allows them to open and close.

histology The scientific study of tissue structure in plants, animals, or nonliving matter.

holdfast A rootlike organ of some benthic alga and seaweeds that anchors them to the seafloor.

holoplankton Planktonic organism living its entire life cycle floating in the water column as a permanent resident of the plankton community.

homoiotherms Warm-blooded animals, such as mammals and birds, whose metabolism regulates their body temperature. This is opposed to poikilotherms, which are cold-blooded, having a body temperature that varies with their surrounding climate, such as fish and reptiles.

Hormuz, Strait of The elbow of water connecting the Gulf of Oman with the Persian Gulf, centered at about 27° N and 56° E between Iran, to the north, and the Arabian Peninsula, to the south.

horologist An expert in the science of measuring time. *See also* HARRISON, JOHN.

horse latitudes Regions of calm, mild winds lying between the belts of the trade winds and the westerlies at about latitude 30° north and south. For sailing vessels, the lack of wind sometimes caused long delays. The term may have derived from the many horses that died on ships as a result of a depletion of the ship's stores, or from a Spanish sailing term comparing the area's unpredictable wind to a mare's erratic disposition.

host An organism in or on which a PARASITE lives.

Hudson Bay Centered at about 60° N and 85° W, this is a large inland body of water located in the heart of northeastern Canada and named for the English navigator HENRY HUDSON, who first explored it in 1610. Hudson's Bay Company set up numerous trading posts along the river outlets. Many of these ports still exist and have been operating since 1670.

Hudson Strait An arm of water in northeastern Canada that connects the great Hudson Bay, to the west, with the LABRADOR SEA, to the east, and ultimately to the Atlantic Ocean. It is centered at about 62° N and 72° W and separates Baffin Island to the north from Quebec to the south.

hull The main body of a vessel.

Humboldt Current *See* PERU CURRENT.

humidity A measured percentage of airborne water vapor, which varies.

hydra Aquatic animal, about 0.1 to one inch (0.25–2.5 cm) in length, belonging to the same class as the PORTUGUESE MAN-OF-WAR. Their hollow bodies are double-layered and cylindrical, having a "foot" on one end and six to 10 tentacles surrounding its mouth at the other. Of the multicellular animals, they are among the simplest in structure. They move by gliding along on the foot as a snail does or by somersaulting. Hydras belong to the order Hydroida, class Hydrozoa, phylum Cnidaria (Coelenterata), kingdom Animalia.

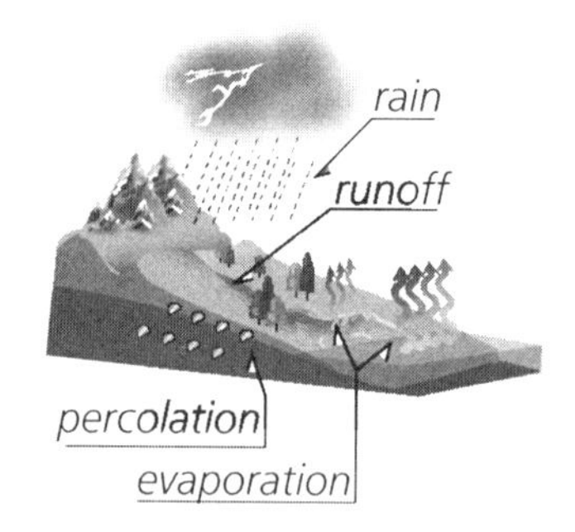

hydrological cycle

hydrography The physical study of ocean and lake features. Hydrography is the aquatic version of geography (study of physical land features).

hydrological cycle The exchange of water among the oceans, atmosphere, and land through evaporation, rain, land runoff such as rivers and streams, and groundwater percolation.

hydrology The scientific study of Earth's water: its distribution, chemical and physical properties, and reaction with the environment, including living things.

hydrophilic Liking or attracted by water.

hydrophobic Disliking or repelled by water.

hydrophytes Plants that thrive in fresh open water or very soggy ground.

hydrothermal vents
(Courtesy of NOAA/NURP)

hydrosphere Constitutes all the water of the world in its liquid form, meaning the lakes, seas, oceans, rivers, and so on. The term is used to distinguish the watery parts of the Earth from the rocky part (lithosphere), the gaseous part (atmosphere), and the living biological part (biosphere).

hydrothermal vents Deep-sea, mineral-rich hot springs found mainly around areas of rapid SEAFLOOR SPREADING, most of them in the Pacific Ocean. Some vents, called black smokers, belch opaque volcanic clouds. Water from these vents can reach up to 660° F (350° C).
See also BALLARD, ROBERT D.

hypersaline Liquid with a SALINITY content of 40 percent or higher.

IAPSO *(International Association for the Physical Sciences of the Ocean)* One of seven associations of the International Union of Geodesy and Geophysics.
See also Useful websites in the Appendixes.

Iberian Basin A basin of the North Atlantic Ocean centered at about 40° N and 20° W that includes the Tagus and Horseshoe Abyssal Plains. The European Basin borders it to the north, the CANARY BASIN to the south, Europe to the east, and the Mid-Atlantic Ridge to the west. The point of greatest depth is roughly 18,000 feet (5,486 m).

iceberg In the ocean, a free-floating ice mass calved (sliced off) from a glacier or ice shelf. An iceberg can be any size, even miles, in diameter.

ICES *(International Council for the Exploration of the Sea)* A forum in which scientists of 19 member countries meet to exchange information about the seas and to coordinate and promote marine research.
See also Useful websites in the Appendixes.

ice shelf A great mass of permanent ice attached to a continent but floating out across the sea like a fringing shelf. The ice sheet (that which lies across the land) of Antarctica is about 11 percent ice shelf, and its largest, the Ross Ice Shelf, is roughly the size of France.

ichthyology A branch of zoology that studies fishes. The prefix *ichthyo-* (ik' - thee-o) is used in forming compound words such as *ichthyophagist*, one who subsists on fish, or *ichthyoid*, meaning "fishlike" or "any fishlike vertebrate."

ichthyoplankton In the water column, planktonic fish larvae and suspended fish eggs. Ocean samplings of ichthyoplankton, to assess the adult spawning biomass distributions of particular fish, began in the late 1800s.

IMI *(Irish Marine Institute)* A national agency of Ireland dedicated to the promotion of and provision of services for marine research and development.

Indian Ocean Earth's third largest of the five major oceans, ranking in size behind the Pacific, then the Atlantic. Its surface area covers approximately 26,470,000 square miles (68,556,000 km^2), bordered by southern Asia, Antarctica, eastern Africa, and southeastern Australia. It constitutes about 20 percent of the world's total ocean area, and its origins date back to the Mesozoic era.
See also ARCTIC OCEAN, ATLANTIC OCEAN, PACIFIC OCEAN, *and* SOUTHERN OCEAN.

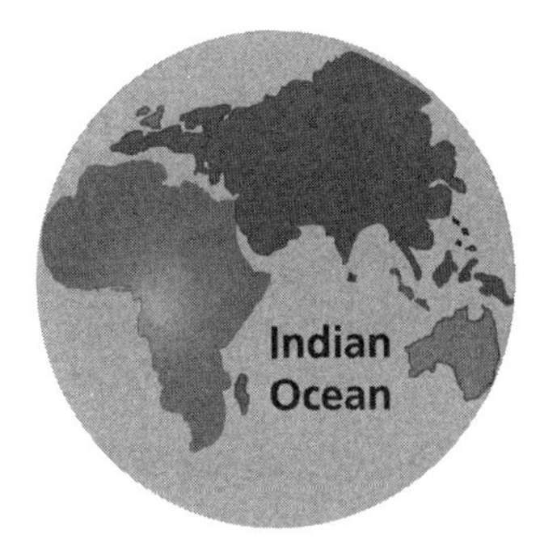

Indian Ocean

infauna Animals that live buried in the sediment.

infralittoral zone Also called sublittoral zone, is the lowest part of the shore, exposed only at the very lowest tides. Examples of life in this zone are a wide variety of algae, surfgrass, and sea palm, which in turn provide shelter for an array of other organisms—sea urchins, sea stars, limpets, anemones, snails, periwinkles, nudibranchs, kelp fish, and sponges, to name a few.

Inland Sea 1) The name given to a shallow arm of the Pacific Ocean linked to the Sea of Japan by a narrow channel. The Inland Sea occupies about 3,672 square miles (9,510 km^2), lying in South Japan among Honshu, Shikoku, and Kyushu Islands. Scattered throughout lie 950 smaller islands. 2) Any sea surrounded by land, connected to the ocean by one or more narrow straits. Examples are the Baltic, Red, and Black Seas.
See also EPEIRIC SEA; EPICONTINENTAL SEA.

inshore zone The area of shore extending seaward from the foreshore to just beyond the breaker zone.
See also NEARSHORE ZONE.

interbasin exchange The active exchange of water between oceanic basins.

interface The boundary that separates layers maintaining different properties, such as density, temperature, and chemical composition.

interlittoral zone Also called intertidal zone, the area of the shore between the highest and lowest points reached by the daily tides.

intermediate water Polar water masses that sink at convergence zones to a depth of between 2,600 and 3,300 feet (792–1,006 m) before spreading as they begin to flow toward the equator.

international date line An imaginary line on Earth that follows the 180th meridian (longitude), with a few meanderings to accommodate the fringes of countries on the globe. The IDL (international date line) is exactly opposite the PRIME MERIDIAN, which runs through Greenwich, England (0° longitude). The date on the east side of the IDL is one day earlier than that just to the west, although in some places, such as between eastern Siberia and the tip of Alaska's Aleutian chain, the landforms are very close to one another.

interstitial Pertains to the cavities and channels formed by the spaces between grains in a sediment (interstitial space).

intertidal zone *See* INTERLITTORAL ZONE.

Intertropical Convergence Boundary air masses between the northeast and southeast trade winds that originate in the Northern and Southern Hemispheres then converge to produce generally wet and cloudy weather. The mean position is somewhat north of the equator.

invertebrate Animals lacking a backbone, of which there are more than a million known species. Examples are JELLYFISH, SPONGES, CORALS, SEA ANEMONES, WORMS, MOLLUSKS, and ARTHROPODS.

Ionian Sea Poised beneath the "boot" of Italy, this is a body of water belonging to the MEDITERRANEAN SEA centered at about 38° N and

17° E. It connects the ADRIATIC SEA, lying north, to the Mediterranean proper, lying south, and is bordered by Italy and Sicily to the west and Greece to the east.

Irish Sea Centered at about 54° N and 5° W, this is a shallow sea belonging to the waters of the North Atlantic Ocean. It separates Ireland, to the west, from Great Britain, to the east, and connects the United Kingdom's Saint George's Channel on the south with the North Channel on the north.

Irminger Current A current of the North Atlantic Ocean setting westward off the southwest coast of Iceland. A branch of the NORTH ATLANTIC CURRENT, it meets the EAST GREENLAND CURRENT as it flows south along the Greenland continental slope.

Irminger Sea An area of water recognized only by some geographers that may not be found on many charts. Its boundaries are marked more by hydrological, than geological masses. It lies roughly between the west coast of Iceland and the east coast of Greenland with the LABRADOR SEA in its southwest corner and the GREENLAND SEA to the northeast.

island A landmass smaller than a continent and surrounded by water, of which there are four kinds: a continental island, which is an island that was once attached to a continent; a volcanic island, made up of lava built upward from the ocean floor; a coral island, consisting primarily of coral materials; and a barrier island, consisting of sediment built up along a shoreline and acting as a natural BREAKWATER.

island arc A curving range of volcanic islands that form along an oceanic trench, such as Alaska's Aleutian chain and those of Japan.

isohaline Constant or uniform in salinity.

isopycnic Constant or uniform in density.

isothermobath A line drawn through a vertical section of the ocean connecting points of equal temperature.

Jan Mayen Current A current of the GREENLAND SEA setting eastward past the north side of Jan Mayen Island, Norway, which is located between the Arctic and Atlantic Oceans.

Japan, Sea of Centered at about 39° N and 134° E, this is a large marginal sea belonging to the North Pacific Ocean. It separates Japan, to the east, from the Koreas and Russia, to the west, and connects to the East China Sea on the south by way of the Korea Strait.

Japan Trench An oceanic trench of the North Pacific centered at about 37° N and 143° E, just off the east coast of the Japanese islands. From it, the KURIL TRENCH snakes northward and the Izu, Bonin, and MARIANA TRENCHES southward. Its deepest point is roughly 28,000 feet (8,534 m).

Java Sea A body of water centered at about 5° S and 112° E amid the Indonesian islands. The island of Java borders it to the south, Sumatra to the west, the BALI and FLORES SEAS to the east, and Borneo to the north.

jellyfish Semitransparent, bell-shaped pelagic coelenterate, often with long tentacles bearing cnidocytes (stinging cells). It is not a fish.

jetsam Floating debris washed ashore, often from ships. Compare to FLOTSAM.

jet stream An atmospheric phenomenon that affects ocean currents and storm tracts.

jetty A structure, usually rock, similar to a BREAKWATER and designed primarily to direct current and inhibit the encroachment of silt into navigable channels.

jib A triangular sail that runs from the foremast head to the jib boom, or in smaller craft to the bowsprit or bow. As a verb, *to jib,* means to swing the jib sail or jib boom around to change course (called tacking).

Jim suit A one-atmosphere armored dive suit first developed and demonstrated in 1930 by the engineer and inventor Salim "Joseph" Peress. Peress's assistant, Jim Jarret, took the prototype, named the Tritonia Diving Suit, to a depth of 450 feet (137 m) in September of that year. The modern Jim suit, named for Jarret, can be used for diving to about 1,970 feet (600 m) below the surface, protecting the diver from the effects of water pressure at that depth. The WASP is a Jim suit with thrusters.

John Dory fish *See* DORY FISH.

JOIDES *(Joint Oceanographic Institutions for Deep Earth Sampling)* A program based in Miami, Florida, that obtains core samples of deep ocean sediments.

Juan de Fuca Ridge An undersea volcanic ridge centered at about 49° N and 128° W off the western coast of Canada in the North Pacific Ocean. North America borders it to the east, the undersea Mendocino Fracture Zone to the south, the Queen Charlotte Islands to the north, and the undersea Tufts Plain to the west.

Juan de Fuca Strait A narrow international passage centered at about 48° N and 124° W between the state of Washington in the northwestern

United States and Vancouver Island, British Columbia, Canada. It is named for the Greek explorer Juan de Fuca, who sailed for Spain and explored the Strait in 1592.

junk Chinese sailing vessel used for centuries as a warship, fishing vessel, and exploration and trade vessel. A junk has a squared-off bow and stern, is built higher in the stern, and is rigged on two or three masts with large distinctive lugsails ribbed with bamboo. It has long rudders and a flat bottom, which increases stability in open waters.

junk (Photograph by George Saxton/Courtesy of NOAA)

kamal A navigating device developed by the ancient Arabs that allowed them to determine latitude by gauging the height of Polaris, the north star, above the horizon.

Kamchatka Current A cold current setting southward from the BERING SEA past Russia's Kamchatka Peninsula. It combines with the Alaskan Stream to form the OYASHIO CURRENT.

Kara Sea A marginal sea of the ARCTIC OCEAN centered at about 75° N and 70° E. The islands of Novaya Zemlya border it to the west and Severnaya Zemlya to the east, Russia to the south, and the Arctic Ocean proper to the north.

Kattegat An ocean strait between Sweden and Denmark centered at about 57° N and 11° E. It connects, westward, with the NORTH SEA through the SKAGERRAK, and with the BALTIC SEA, eastward.

keel The absolute bottom, centerline of a boat running fore and aft.

kelp A family of large brown seaweeds that forest the oceans. The largest grow to 100 feet (30 m) or more in length.

Kerguelén Plateau Centered at about 55° S and 73° E, this is a broad oceanic ridge of the southern Indian Ocean. It reaches southward from the Kerguelén Islands toward Antarctica and is large enough to hinder the flow of the ANTARCTIC CURRENT.

Kermadec Trench A deep oceanic trough centered at about 32° S and 177° W near the international date line in the South Pacific Ocean. The undersea Tonga Trench borders it to the north, the New Zealand islands to the south, the undersea South Fiji Basin to the west, and the SOUTHWEST PACIFIC BASIN to the east. It reaches a maximal depth of roughly 33,000 feet (10,058 m).

ketch A modern two-masted sailing vessel, exclusively used as a pleasure boat today. Early square-rigged ketches were used as coastal fishing and trading ships. Later they were also adopted into the navy as bombing vessel, with mortar-fired missiles used as bombs.

key Or cay, a chain of low, flat islands, islets, and reefs consisting largely of limestone and coral. One example, the Florida Keys, were devastated in a hurricane in 1935; today many of the islands are popular with vacationers.

keystone species A species that through its uniquely important effects, whatever they may be, helps maintain the equilibrium of an environment. The removal of such would result in dramatic, usually detrimental, changes in the ecological system.

killer whale The largest dolphin. *See* ORCA.

kingdom In TAXONOMY, based only on the very basic of similarities and common ancestry, the highest classification into which organisms are grouped. The organisms of earth are classified in five basic kingdoms: Monera, Fungi, Plantae, Protozoa, and Animalia. "Beneath" each of those kingdoms, the general taxonomic divisions are phylum, class, order, family, genus, and species.

Kinorhyncha Phylum of small marine wormlike invertebrates, within the kingdom Animalia.

Korea Current A warm surface current of the Sea of Japan that moves north and then east to become the Tsugaru and Soya Currents.

Korea Strait Centered at about 34° N and 130° E, this is a body of water connecting the eastern side of the East China Sea with the southern end of the Sea of Japan. It separates Japan to the southeast from South Korea to the northwest.

knot A unit of speed used by vessels in water equal to one nautical mile (6,076 feet [1,852 m]) per hour. The term has derived from the early form of determining a ship's speed. While the ship was under way, a long line with a small wooden board at the end and knots tied every 47 feet three inches (14.4 m) would be paid into the sea for an interval of 28 seconds. If within that time, for example, three knots passed through their hands, they knew they were traveling at three knots, or three NAUTICAL MILES per hour.

krill Pelagic, shrimplike crustaceans that form the mainstay diet for large marine animals such as many species of whales and Antarctic and sub-antarctic penguins. There are around 90 species, ranging from 0.3 to three inches (8–80 mm). Krill make up the order Euphausiacea, sub-class Malacostraca, class Crustacea, phylum Arthropoda, kingdom Animalia.

Kuril Trench A deep oceanic trough centered at about 48° N and 156° E off the east coast of Russia. It connects with the ALEUTIAN TRENCH, north, and the JAPAN TRENCH, south, and is bounded by the NORTHWEST PACIFIC BASIN to the east and the Kuril Basin and Sea of OKHOTSK to the west. Its point of greatest depth is roughly 34,587 feet (10,542 m).

Kuroshio Extension A current in the North Pacific Ocean heading eastward from about longitude 145° E to about 160° E, where it continues the flow of the Cherish Extension to the NORTH PACIFIC CURRENT.

kymatology The science that studies waves and WAVE motion.

Labrador Basin An oceanic basin centered at about 53° N and 45° W in the North Atlantic. It lies along the path of the Northwest Atlantic Mid-Ocean Canyon and is bounded by Newfoundland, Canada, to the west; the undersea Charlie-Gibbs Fracture Zone to the east; Greenland to the north; and the undersea Flemish Cap to the south. The point of greatest depth is roughly 11,500 feet (3,505 m).

Labrador Current A cold, low-salinity current setting southward from the Arctic and Davis Strait along the coasts of Baffin Island, Labrador, and Newfoundland.

Labrador Sea A body of water belonging to the North Atlantic Ocean centered at about 57° N and 55° W. Newfoundland borders it to the west, Greenland and the Davis Strait to the north; elsewhere it is open to the North Atlantic.

Laccadive Sea A marginal sea of the North Indian Ocean centered at about 5° N and 75° E between the western coast of India and the waters of the ARABIAN SEA.

lagoon Typically a shallow body of marine water harbored from the open sea by breakwaters of natural sandy islands, dots of coral reefs, or various other types of semisubmerged features.

lamella A thin layer, membrane, or platelike formation of a) gills in fish or bivalve mollusks or b) fine, hairlike formations affixed to the bills of certain birds, which enable food particles to be filtered from water.

lampshell A class of small marine bivalves once classed as a mollusk but reclassed because its shell lies ventrally and dorsally, as opposed to the shell of a mollusk, which runs left and right. Also, the lampshell has armlike appendages at each side of its mouth, which as a result resembles an ancient oil lamp with protruding wicks and inspired the name lampshell. Lampshells—also called brachiopods—live at moderate depths and usually measure less than one inch (2.5 cm) across. These animals make up the phylum Brachiopoda, kingdom

Animalia. Hinged brachiopods make up the class Articulata, and those without hinges make up the class Inarticulata.

lanternfish A bioluminescent fish whose photophores generally develop on their undersides, glowing just enough to match the light levels of the sea around them. Their backs are dark, blending with the deep water of the mesopelagic zone. At night they rise toward the surface. Adults grow to about six inches (15 cm) in length. There are about 230 species of lanternfish and over 1,000 species of bioluminescent fish. Lanternfish belong to the family Myctophidae, order Myctophiformes, superorder Scopelomorpha, subclass Neopterygii, subphylum Vertebrata, phylum Chordata, kingdom Animalia.

Laptev Sea A marginal sea of the Arctic Ocean centered at about 75° N and 128° E. Russia borders it to the south, the Arctic Ocean to the north, the EAST SIBERIAN SEA to the east, and the KARA SEA to the west.

larva The stage of growth in certain types of animals before they reach maturity (plural, larvae).

lateen sail A triangular-shaped sail, originally called a latin sail, for its use in the Mediterranean Sea.

lateral line The sensory system along the sides of a fish that detects disturbances and vibrations that occur within the water.

latitude Location on the Earth's surface measured in degrees north and south of the EQUATOR, with the equator at latitude 0° and the poles at 90°.

Laurasia One of two supercontinental masses that were formed by the breakup of the supercontinent PANGAEA; the other is GONDWANALAND. From it North America, Greenland, Europe, and much of Asia in the Earth's Northern Hemisphere were formed.

See also CONTINENTAL DRIFT; PLATE TECTONICS.

laver Laver, as it is known in England, is a genus of blade-forming red seaweed useful to us as food. The black-red wrapping on sushi rolls is prepared laver. It grows in sheets, collecting at the high water mark in temperate seas. In Japan and North America it is known as nori, in Korea as kim, and in New Zealand as karengo. Laver is the focus of a billion-dollar aquaculture industry in Japan, Korea, and the People's Republic of China.

lazarette On a ship, a storage area in the stern.

lead An opening in pack ice that occurs at weak points as a result of a shift in its drift. The opening serves as a major migratory corridor for marine mammals and sea birds in the arctic spring.

Leeuwin Current A strong current of the Indian Ocean that flows northward along the coast of Australia, defying the force of the prevailing winds because of the strength of the flow coming from the Pacific.

leeward The sheltered side of an object (island, ship, etc.) exposed to wind. This is the opposite of WINDWARD.

lichen Characterized as any group of composite organisms made up of a fungus and an alga living as symbionts. There are about 15,000 known species of lichen, which usually grow in harsh conditions on rocks, bark, or poor soil. Some lichens, such as arctic reindeer moss, can serve as food.

Lichenes A phylum of the kingdom Fungi that encompasses the lichens.

lichenometry A technique used to date the ages of material exposure throughout the world. It uses the growth rate of a given species of lichen as an indicator of the age of the surface on which the lichen is growing.

lighthouse A coastal structure erected as a navigational aid for ships approaching land. It is used to mark and/or warn of submerged rock formations, reefs, sand shoals, and entrances to harbors. The earliest known built were on the MEDITERRANEAN SEA in the seventh century B.C.E. Other ancient versions were simply large bonfires built on the beach. The most famous ancient lighthouse and one of the Seven Wonders of the Ancient World was the Pharos of Alexandria, built in the time of Ptolemy II (309–247 B.C.E.). Today's lighthouses have bright, rotating beacons that can be seen for miles.

Ligurian Sea A marginal sea of the MEDITERRANEAN SEA centered at about 43° N and 9° E. It is bounded mainly by Italy, with the TYRRHENIAN SEA and the island of Corsica to the south and France to the north.

Liman Current A cold water current of the North Pacific Ocean setting southward from the Sea of OKHOTSK toward the Sea of Japan, where it eventually merges with the KOREA CURRENT.

limnology The scientific study of fresh water, as opposed to oceanography, the study of the marine environment.

limpet A small marine GASTROPOD in the shape of a flat cone. Some species grow to about 1.5 inches (3.8 cm) long. Subsisting on algae, they can be found attached to rocks in littoral zones worldwide, surviving exposure during low tide by trapping moisture under their shells. Limpets belong to the order Archaeogastropoda, class Gastropoda, phylum Mollusca, kingdom Animalia.

limpet (Photograph by Scott McCutcheon)

line As a nautical term, aboard ship all ropes are called lines.

lip In a GASTROPOD's shell, it is the edge of the opening.

lithosphere *See* HYDROSPHERE.

lithotomous Pertains to an organism that burrows into rocks, such as the piddock, a rock-boring MOLLUSK that is thought to secrete rock-dissolving acids.

littoral zones Descriptive of the biogeographic regions of an ocean shore. *See* INFRALITTORAL, INTERLITTORAL, MIDLITTORAL, and SUPRALITTORAL ZONES.

load The amount of SEDIMENT held within a body of water. The load sinks and is deposited as sediment when the water flow decreases.

lobster A marine decapod, of which there are three true lobster species, the American, European, and Norway lobsters. True lobsters have the distinguishing pincers, called chelapeds, one a huge crusher claw and the other a smaller biting claw. They grow by molting and exist by scavenging. Lobsters belong to the suborder Reptantia, order Decapoda, subclass Eucarida, class Malcostraca, subphylum Crustacea, phylum Arthropoda, kingdom Animalia.

lodestone Labeled such by the Europeans in about 1200 C.E., when they discovered that an oblong piece of magnetite (a rock found in nature to have a strong MAGNETIC FIELD that attracts ferrous metals), when hung by a string, would swing to point to the north or south. It was then called the leading stone, or lodestone, as it made the creation of the compass possible. It is said that many men were superstitious about the first compasses and would not sail with a captain who employed one. Reason eventually set in, however, and the usefulness of the compass finally won the day.

Lomonosov Ridge An undersea ridge of the Arctic Ocean separating the MARKOV BASIN from the Pole Plain and FRAM BASIN. It is centered directly beneath the North Pole, and its peaks rise to more than 6,600 feet (2,011 m), reaching to within 3,300 feet (1,006 m) of the surface.

London Royal Society *See* ROYAL SOCIETY OF LONDON.

longitude Location on the Earth's surface measured in degrees east and west of Greenwich, London, which is the PRIME MERIDIAN, beginning at 0°. Opposite, at 180° longitude, stretches the INTERNATIONAL DATE LINE.

long-lining A commercial fishing method in which a very long line is set with a few to hundreds of hooks placed regularly down its length. The line is then towed from the stern and can be fished on the bottom for cod or halibut, or near the surface for tuna and billfish.

longship A rather poorly designed Viking galley ship, usually propelled by many oarsmen but capable of grasping the wind with a square sail on a single mast. The longship was popular between 600 and 1100 C.E. as a trading, raiding, or emigrant vessel, used in seasons of fair weather and put ashore for the winter.

longshore bar A sandbar running generally parallel to a coast that may or may not become exposed at low tide. There may be one or many running alongside one another, extending down the continental slope.

longshore current A current running parallel to the shore, chiefly within the SURF ZONE. It is caused by excess water transported to the zone by the wind-driven waves. Longshore currents feed into RIP CURRENTS.

longshore drift Also called beach drift or littoral drift, it is the drifting of marine sediment in parallel patterns, which occurs when waves BREAK at an angle across the shore. The result is a zigzag pattern along the beach.

longshore trough A long depression running generally parallel to a coast that may or may not become exposed at low tide. As with longshore bars, there may be one or several.

Loop Current A western boundary current deriving from the Mediterranean Current that arrives through the Yucatán Strait and travels into the Gulf of Mexico, where it becomes unstable. Parts whirl off as eddies into the gulf while the rest leaves through the Straits of Florida.

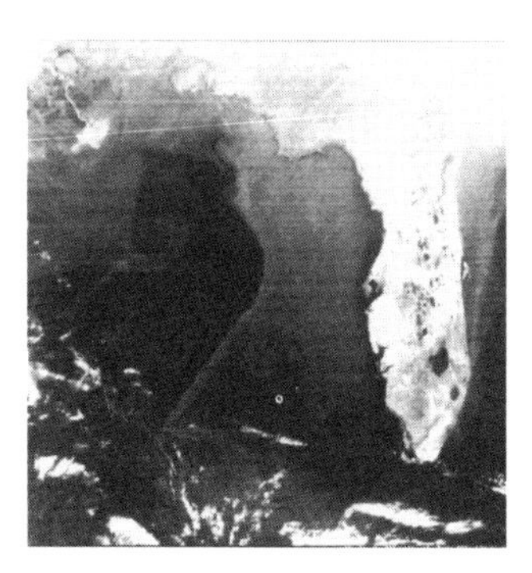

Loop Current (Courtesy of NOAA)

Lord Howe Rise An oceanic ridge of the South Pacific Ocean centered at about 35° S and 165° E. From New Zealand it reaches northward to the CORAL SEA. Australia and the undersea Tasman Plain lie to the west, and the undersea New Caledonia Basin lies directly east.

Loricifera Phylum of spiny, oval-shaped, microscopic interstitial species in shelly gravel. These fall within the kingdom Animalia.

low tide *See* EBB TIDE; TIDES.

low water The minimal height achieved by a falling tide.

lubber's line On a compass, a mark or line that, when properly installed, indicates the direction forward parallel to a ship's keel.

lysocline In ocean SEDIMENT, the upper portion of the transition zone between the supersaturated (shallow) and undersaturated (deep) sediments.

mackerel One of many species of fish belonging to the TUNA family. Mackerels live in EPIPELAGIC ZONES, roaming the ocean in schools. They are fast and strong, great fun to fish for sport, and good to eat. Basic physical features are wavy dark stripes on their backs, silver bellies, five or six finlets behind both the dorsal and anal fins, pointed head, thick, narrow body, and large mouth. Mackerels belong to the family Scombridae, order Perciformes, class Actinopterygii, subphylum Vertebrata, phylum Chordata, kingdom Animalia.

macrophytes Any large marine plants, such as mangroves and sea grasses.

macroplankton Large plankton between .78–7.8 inches (2.0 and 20 cm); examples are some CTENOPHORES.

Madagascar and Mascarene Basins Basins in the midwestern Indian Ocean centered between latitudes 10° S and 30° S at about 55° E off the eastern coast of Madagascar. Points of greatest depth average about 19,580 feet (5,968 m) in the southerly positioned Madagascar Basin and 17,050 feet (5,197 m) in the northerly positioned Mascarene Basin.

Magellan, Straits of Called Estrecho de Magallãnes, it is an elbowlike passage at the tip of South America with access to both the Atlantic and Pacific Oceans through Chile. The bend in the "elbow" is centered at about 55° S and 71° W. The channel is named for its discoverer, FERDINAND MAGELLAN.

magnetic anomaly A point or region where magnetic field strength and/or direction differs from that of the surrounding area. The strips of differing magnetic field strength found paralleling midoceanic ridges are an example of a magnetic anomaly. These differences are a record of new basalt formations and demonstrate that ancient Earth once possessed a weaker magnetic field. Through the phenomenon of SEAFLOOR SPREADING and thus the introduction of new crustal material containing iron, the Earth's magnetic field has become anomalous in places. This phenomenon also occurs with new rock formations on land. The history of the breakup of PANGAEA and the movements of the continents can be reconstructed by tracking the magnetic anomalies on the seafloor.

See also VINE, FRED.

magnetic declination The difference between MAGNETIC NORTH (or south) and GEOGRAPHIC NORTH (or south).

magnetic field The area surrounding a magnetic body or a current-carrying body in which magnetic forces can be detected.

magnetic north The direction indicated by the north-seeking end of a magnetic compass needle.
See also GEOGRAPHIC NORTH; MAGNETIC DECLINATION.

magnetometer A device that can detect and/or measure a magnetic field or the presence of a metallic object. Such a device may be used to detect magnetic anomalies within the seafloor.

mainmast On a ship, the main center support, or SPAR, to which sails are rigged.

Maldives A chain of nearly 2,000 low-lying coral atolls located in the northern Indian Ocean. They lie between the equator and 10° N at about 73° E, southwestward from the southern tip of India. The Maldives are part of the Chagos-Laccadive Ridge, a large group of submerged coral-limestone banks.

Malvinas Current A cold-water current of the South Atlantic Ocean that sets northward past the Falkland Islands as a looping extension of the ANTARCTIC CURRENT.

mammal Any class of endothermic (warm-blooded) vertebrate animals whose young are born live and nurtured on milk from the female. Mammals possess body hair in some degree or another and breath air. Dolphins, sea cows, walruses, sea otters, seals, and whales are all mammals, as are humans. All mammals are members of the class Mammalia, subphylum Vertebrata, phylum Chordata, kingdom Animalia.

manatee *See* SIRENIAN.

mandible The lower jawbone, or either the upper or the lower part of a bird's bill.

mangroves Salt-tolerant trees and shrubs that flourish throughout the intertidal zones of tropical and subtropical regions.

mantle 1) A lobe or pair of lobes that lines the body wall of a shelled mollusk that bears the shell-secreting glands. 2) Inside the Earth, the area that extends between the crust and the core.

map The representation of a celestial body (such as the Earth) or attributes of that body in whole or in part drawn to scale, usually on a flat plane.

map projections Mathematical drawings showing the curved surface of a celestial body, such as the Earth, on a flat plane where direction, shape, or area is represented accurately, but distortion in large regions cannot be prevented.

marginal ice zone (MIZ) An area of sea only partially covered by dispersed ice floes, located between the solid sea ice and open water. This zone typically composes a substantial portion of the total sea area covered by ice. In these zones, such as the Barents Sea, MIZ studies of the interactions among air, sea, and ice are constantly being conducted to provide information such as that used in the study of the effect of greenhouse gases.

marginal sea A partially enclosed body of water bordering a large continent. The Caribbean and Mediterranean are marginal Atlantic seas; the Sea of Japan and the South China Sea are both marginal seas of the Pacific. In the Indian Ocean, marginal seas are the Timor and the Arafura.

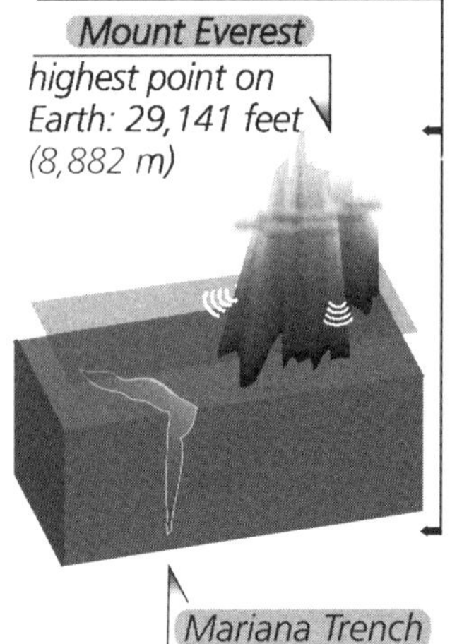

Mariana Trench

Mariana Trench A crushingly deep, sickle-shaped trough in the midwestern Pacific Ocean, known at the time of this writing as the deepest point on the Earth. It extends between 10° N and 20° N, centered at about 145° W. Its deepest known point, about 35,827 feet (10,919 m), is called the Challenger Deep, named after the research vessel and expedition during which it was discovered. Though many new life-forms have been discovered there, only a small percentage of this cleft has been explored, and it merely lies in wait for technology, funding, and human desire to reveal more of its secrets.

mariculture Cultivation of marine organisms under controlled conditions.

marine archaeology The systematic sighting, study, mapping, and sifting of artifacts from oceanic bottom material with the objective of uncovering historical objects that will further enrich our knowledge.

marine biology A major branch of oceanography, the beginnings of which date to the early 19th century. Essentially it is the intimate study of life within the ocean, both animal and plant, mainly of their habituation to each other. It also deals with issues that affect marine life such as ocean currents, waves, temperature, pressure, water chemical, characteristics, and light intensity.

marine oils Oils derived from fish with a high content of fatty acids. Sardine and herring oils can be prepared as nondrying oils for use in paints and ink. Cod liver oil, from the codfish, was long used as a dietary supplement for its richness in vitamins A and D. Spermaceti is a waxy, tasteless, odorless substance separated from the oil of sperm whales that is used in cosmetics, ointments, and good-quality candles.

marine palynology The study of pollen deposits in marine sediment.

marine sediments Organic or mineral-based materials that have settled to the ocean floor through the natural process of death or after being deposited in water by wind or erosion.

marine snow Oceanic particles that fall to the seafloor and help keep clean the surface layers of the ocean. Marine snow can consist of dead PHYTOPLANKTON, ZOOPLANKTON feces, mucous feeding webs of some zooplankton, DETRITAL, or other types of living organisms visible to the naked eye. Organic marine snow is a major food source for organisms living in deeper abyssal waters.

Markov Basin One of four major basins of the Arctic seafloor beneath the ARCTIC OCEAN; the others are the CANADA, FRAM, and NANSEN BASINS. It is centered between the Canada and Fram Basins on the eastern side of the North Pole; the point of greatest depth averages roughly 13,150 feet (4,008 m).

marlin A type of food and sport fish related to the sailfish and spearfish (billfishes), notable for its long, swordlike nose. Marlins are tropical, deep-sea dwelling carnivores, averaging between 50 and 400 pounds (23–181 kg), though they can grow much larger. There are both Atlantic (white marlin) and Pacific (striped marlin) types, but blue and black marlins are found in both these oceans. Marlins belong to the family Istiophoridae, order Perciformes, class Actinopterygii, subphylum Vertebrata, phylum Chordata, kingdom Animalia.

Marmara Sea A body of water important because it allows passage between the BLACK and MEDITERRANEAN SEAS. It is centered at about 41° N and 28° E and, except for the tiny channels of Bosporus (east to the Black Sea) and Dardanelles (west to the Adriatic), it is encircled entirely by western Turkey.

masl Abbreviation of meters above sea level.

mass stranding The beaching of a marine species that normally lives within the water; beaching of cetaceans draws most attention. On September 12, 1993, along the southern coast of England near Torbay, a gale from an unusual direction caused up to 10-foot (3-m) swells that swept over rocks and caused waves as tall as 30 feet (9 m) to break onto coastal areas normally more protected from weather. The result was the mass stranding of many marine animals, mostly mollusks, which were carried to shore. In October 2000, a pod of six white-sided dolphins stranded themselves on the beach at Brora, on the east coast of Scotland. There is no single accepted answer why whales strand, but there are many theories.

maximum sustained yield Maximum number of the yearly harvest of a species without depleting the stock, which ensures that the remaining surviving stock is large enough to reproduce naturally and replace the lost numbers. *See* OVERFISHING.

mean Something with a condition that is midway between extremes; for example, mean high tide is the average expected height reached by a high tide without considering whether it might be extremely high or low for that particular high tide.

mechanoreceptor A sensory reception system in animals that responds to mechanical influences, such as touch and sound. The shrimplike copepod employs both mechano- and CHEMORECEPTORS to help it detect and escape from potential predators.

Mediterranean Sea A large marginal sea between Europe and Africa connected to the waters of the Atlantic Ocean by way of the Straits of Gibraltar. It is centered at about 38° N between 5° W and 35° E, a surface area covering about 1,112,000 square miles (2,880,000 km^2).

medusa Biological name for a JELLYFISH, in particular the cnidarian (coelenterate) animal in the form of a bell or dome with the mouth located at the center bottom between the tentacles. Large jellyfish belong to the class Scyphozoa, small jellyfish to the class Hydrozoa, phylum Cnidaria, kingdom Animalia.

megaplankton Large planktonic organism whose length falls between 7.8–78 inches (20 and 200 cm), for example, the PORTUGUESE MAN-OF-WAR.

meiobenthos Small organism living buried in marine sediments.

meiofauna Small animals, usually invertebrates, that exist interstitially (between sand grains).

menhaden Members of the herring family, these fish are very silvery and have a distinctive black spot behind the gill and various smaller spots along the sides. Menhaden spawn in the sea and mature in less saline estuaries. They are rich in oil and are caught for use as bait, fertilizer, and animal feed but are not considered food fish. Atlantic types grow to about 15 inches (38 cm) in length. Menhaden belong to the family Clupeidae, order Clupeiformes, class Actinopterygii, subphylum Vertebrata, phylum Chordata, kingdom Animalia.

Mercator projection Named for the Flemish cartographer GERARDUS MERCATOR, who first graphed it in 1569, it is a cylindrical map projection that accurately shows direction but greatly distorts continental scale in high latitudes. During the 17th and 18th centuries, it was the standard for maritime mapping.

mercury A chemical element that when formed in water as a soluble compound, such as methylmercury, may be incorporated into fish populations and in turn poison whatever consumes the fish, including humans. Mercury, like lead, is a very heavy element; it poisons because once ingested it is too heavy for the body to eliminate. *See also* 2000.

meridian A name for the lines of LONGITUDE running pole to pole on the Earth, intersecting the equator.

mermaid's purse Name sometimes used for the leathery egg case of some cartilaginous fishes, such as rays and skates.

meroplankton Floating planktonic organism that spends only part of its life in the WATER COLUMN as eggs and larvae and that, as an adult, belongs to the nekton and benthos.

mesoderm In an embryo, the central layer of cells between the ECTODERM and the ENDODERM, which gives rise to the tissue growth of the muscles, vascular system, and connective tissues.

mesopelagic zone The area of open ocean extending from around 700 to 3,300 feet (213–1,006 m) beneath the EPIPELAGIC ZONE. This is the beginning of the shadowy APHOTIC ZONE, where creatures such as deep-sea eels, deep-sea swallowers, and stalkeyed fish silently haunt the depths.

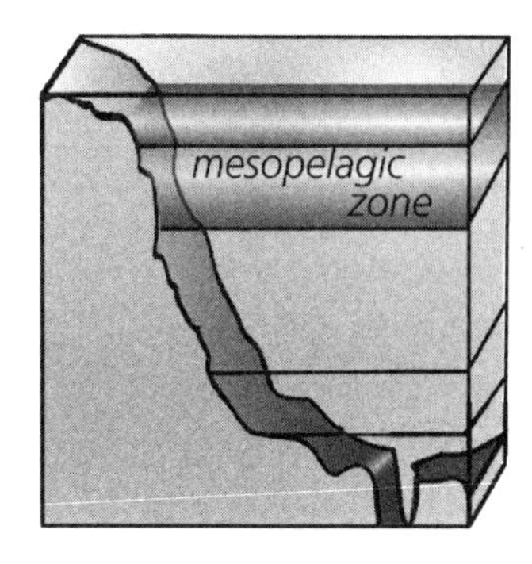

mesopelagic zone

mesoplankton Planktonic organism of 0.00787–0.78 inch (0.2–20 mm); an example is a large DIATOM.

mesotrophic Descriptive of waters containing moderate levels of nutrients, as opposed to EUTROPHIC waters, which contain a high concentration of nutrients, and OLIGOTROPHIC waters, which contain lower concentrations of nutrients.

Messina, Strait of Called Stretto di Messina, it is a finger of water separating Sicily from Italy that provides, to experienced navigators, access between the IONIAN and TYRRHENIAN SEAS. It is centered at about 38° N and 15.5° E.

metamorphosis Abrupt developmental change that occurs in many types of animals in which the adult form differs greatly from the juvenile form.

Metaphyta The kingdom of multicelled plants.

metazoan A multicelled animal.

meteorology The scientific study of the atmosphere, especially weather. Meteorologists collect meteorological data in order to predict future weather conditions.

Mexico Basin Centered at about 24° N and 93° W, this is a western section of seafloor in the Gulf of Mexico that has a long exploration history for the active recovery of natural gas and some deposits of oil. The Sigsbee Deep is the basin's deepest point, at about 12,714 feet (3,875 m).

Mexico, Gulf of A large body of water belonging to the Atlantic Ocean indenting the southeastern portion of the North American continent. Centered at about 24° N and 90° W, the gulf is bordered by the coasts of Mexico (west) and the United States (north); the island of Cuba constricts the mouth to the east. It occupies an area of roughly 620,000 square miles (1,600,000 km^2).

microbenthos Benthic animals that are very small, such as many species of PROTOZOANS.

microorganisms Single-celled organisms visible only through a microscope.

microplankton Plankton not easily seen without a microscope composed of individuals less than 0.039 inch (1 mm) in size, 20–200 μm, but large enough to be retained in a small-mesh plankton net.

Microspora and Sporozoa Phyla of the parasitic protozoans. These fall within the kingdom Protozoa.

Mid-Indian Basin A basin centered in the middle of the Indian Ocean extending through latitudes 30° S to 7° N at about 80° E. The undersea Mid-Indian Ridge borders it to the west, the Ninetyeast Ridge to the east, the Southeast Indian Ridge to the south, and Sri Lanka and the Bay of Bengal to the north. The point of greatest depth is roughly 17,800 feet (5,425 m).

Mid-Japan Sea Current Also known as the Maritime Province Current, it is a cold, slow-moving, and prominent circulation feature of the Sea of Japan.

midlittoral zone The shore area between the SUPRALITTORAL ZONE and the INFRALITTORAL ZONE that experiences regular submergence and exposure twice daily from rising and falling tides. Examples of the wide array of life in this zone are algae, barnacles, mussels, sea urchins, limpets, sea stars, snails, periwinkles, anemones, crabs, chitons, tunicates, and sea lettuce.

midoceanic ridge A chain of undersea mountains that are not part of a continent. *See* SPREADING RIDGE.

migration An animal's or group of animals' seasonal movement from one location to another, normally for the purpose of locating more fertile feeding grounds or reproducing.

milt Secretions, or soft roe, from the reproductive organs of male fish.

mimicry An animal's imitation, through color change, of another animal or object, often to camouflage itself for protection.

Mindanao Current A current of the North Pacific Ocean beginning as the NORTH EQUATORIAL CURRENT and then splitting and moving southward as it encounters the eastern coast of the Philippines.

mixed layer In the ocean, a surface layer of nearly constant temperature of around 150 to 500 feet (46–152 m) in depth caused by the stirring and convection effect of the wind. The depth of the mixed layer can depend on the season: deeper in winter temperatures and shallower in summer.

mizzenmast The aftermost and often the smallest mast of a sailing vessel. *Mizzen* also applies to the sail set on the mast.
See also SPANKER.

Mollusca Phylum of the archaic mollusks, chitons, solenogasters, gastropods, bivalves, tusk shells, and cephalopods. These fall within the kingdom Animalia.

mollusk Invertebrate animal characterized by a soft body, muscular foot, and calcareous shell, with the exception of some species such as the squid and cuttlefish. All mollusks possess a mantle (pallium), which is a skinlike organ that constitutes the animal's shell, and all must maintain moisture within their bodies for survival. Mollusks make up the largest group of water animals. Examples are clams, snails, slugs, mussels, oysters, squids, and octopuses (or octopi).

molt A process by which an animal grows a new body covering by shedding its old skin, hair, scales, feathers, or fur. Crabs, lobsters, and insects shed an exoskeleton in order to grow.

Molucca Sea A sea within the Indonesian islands centered along the equator at about 126° E. Sulawesi island borders it to the west, the Moluccas to the east, the PHILIPPINE SEA to the north, and the BANDA SEA to the south.

Monera A kingdom of simple organisms lacking any kind of distinct cell nucleus, under which fall three phylums: Schizophyta (bacteria), Cyanophyta (blue-green algae), and Archaebacteria (ancient bacteria).
See also BACTERIA; SCHIZOPHYTA.

monoculture Cultivation of only one species of organism in aquaculture or any other biotic system.

Monsoon Current A current of the Indian Ocean that travels east-southeast off India and Sri Lanka during the months of August and September, when it replaces the NORTH EQUATORIAL CURRENT with its wind-driven switch.

morphology The science of studying the structure of living plants and animals and, to a lesser degree, nonliving things. In geology, morphology is the scientific study of the outer formation of rocks.

mortality rate The rate at which a population dies; a prediction of a death rate.

moss animal *See* BRYOZOAN.

Mozambique Current A portion of the AGULHAS CURRENT, from about 10°S to 35°S, flowing south-southwest between the African coast and Madagascar.

MTS *(Marine Technology Society)* An international multidisciplinary society devoted to ocean and marine engineering, the promotion of marine science, and the development of tools for the exploration, study, preservation, and civilized exploitation of the world's oceans.
See also Useful websites in the Appendixes.

mud CLAY and SILT particles of sediment smaller than 0.00236 inch (0.06 mm).

mud flat Low-lying expanse of mud exposed at low tide and covered at high tide. Mud flats can be dangerous to walk on during low tide for even the most experienced. A person or animal can become trapped in the sucking mud and, in extreme circumstances, drown under incoming high tide. *See also* QUICKSAND.

murex Any member of a large group of widely distributed marine gastropods common in tropical seas. The source of the royal color Tyrian purple is this animal, which produces a yellow fluid that turns to purple in sunlight. Murex belong to the class Gastropoda, phylum Mollusca, kingdom Animalia.

mussel A soft-bodied aquatic animal living within a hard shell. There are both marine and freshwater species. The freshwater varieties are found buried in sandbars, marine varieties in sessile beds on hard seafloor objects. The most common is the blue mussel, which has a blue-black shell that is pearly blue on the inside. This animal feeds on plankton and grows to about three inches (7.6 cm) in diameter. Mussels belong to the class Bivalvia, phylum Mollusca, kingdom Animalia.

mutualism A symbiotic association of different organisms in which both benefit in some way or another.

Mycophyta The fungi. Mycophyta is a phylum of the kingdom Fungi

nadir The point of the celestial sphere that is diametrically opposite the ZENITH, directly below the observer.

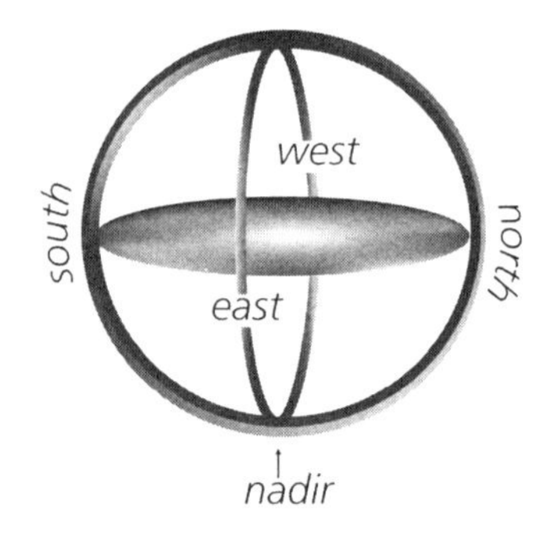

nadir

nanoplankton Plankton smaller than 10 micrometers, 2.0–20 μm, which includes protozoans and phytoplankton, that pass through an ordinary plankton net but can be removed from the water by filter centrifuge.

Nansen Basin One of four major basins of the Arctic seafloor beneath the ARCTIC OCEAN; the others are the CANADA, FRAM, and MARKOV BASINS. It lies between the Nansen Ridge and the continental shelf of Russia. The point of greatest depth is roughly 12,830 feet (3,910 m).

Nansen bottle A type of ocean water sampling bottle developed in 1910 by the Norwegian explorer and scientist FRIDTJOF NANSEN. It is a cylindrical container that extracts samples of seawater to test for temperature and salinity; it is composed of a pair of thermometers and a bottle on a reversible frame. It has generally been replaced by electronic instruments, but some ships still use it today.

Nansen bottle (Photograph by Captain Robert A. Pawlowski/Courtesy of NOAA)

Natal Basin A basin of the southwestern Indian Ocean centered at about 35° S and 40° E between the southeastern coast of Africa and Madagascar. The point of greatest depth is roughly 20,640 feet (6,291 m) near the Mozambique Escarpment.

natural selection A theory made known by the British scientist and naturalist CHARLES DARWIN, stating that organisms that are the fittest, strongest, and most capable of adaptation survive.

nautical mile A unit of measurement equal to 1/60 of one degree, or one minute of arc, of the Earth's circumference (hence the saying "A mile a minute"). At sea, this measurement is used because of its natural relationship to the degrees and minutes by which the longitude and latitude of a vessel's position are measured.

See also KNOT.

nautilus A cephalopod MOLLUSK whose shell is separated into a series of compartments, the largest and most recently formed of which the animal inhabits (the siphuncle). Living in warm waters near the ocean floor, it eats diatoms, shrimp, algae, crabs, and other animals, which it catches with its nearly 100 tentacles that encircle the mouth. Today's nautilus is the only surviving descendant of a now-extinct subclass of nautiloids, animals that lived on the very young Earth. During a period 450 million years ago, the nautilus was one of the largest predators in the sea. The nautilus belongs to the family Nautilidae, order Nautilida, class Cephalopoda, phylum Mollusca, kingdom Animalia.

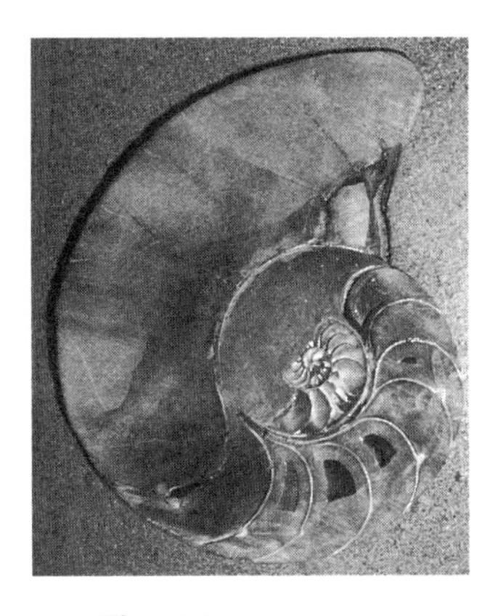

nautilus (Photo courtesy: Royal Tyrrell Museum/ Alberta Community Development)

navigation The art and science applied to conducting a ship, or any other means of transportation, safely from one point to another.

Nazca Ridge A seismic oceanic ridge of the South Pacific Ocean centered at about 20° S and 80° W off the coast of Peru. It runs directly perpendicular from the continent, and its highest peaks reach to about 1,080 feet (329 m) of the ocean surface.

neap tide The smallest range of tides that occur semimonthly as the result of the Moon's being in quadrature, or first and third quarters. The average height of the high waters of the neap tide is called mean high water neaps (MHWN), and the average height of the corresponding low waters is called mean low water neaps (MLWN).

nearshore zone The area of shore extending from the upper limit of a beach to the offshore zone. In terms of describing a beach, traveling seaward the zones consist of BACKSHORE, FORESHORE, and INSHORE ZONES. In terms of describing waves and current, traveling seaward the zones consist of the SWASH, SURF, and BREAKER ZONES.

nekton Strong-swimming pelagic animal capable of swimming against a current, such as adult squid, fish, and marine mammals.

nematocyst The stinging structure normal to cnidarians (COELENTERATES) used for defense, food gathering, and attachment.

Nematoda Phylum of the roundworms, within the kingdom Animalia.

nematodes Invertebrates of the phylum Nematoda possessing a cylindrical body, a conspicuous body cavity, and a complete digestive tract; also known as the roundworms. Roundworms are found worldwide and may be marine, freshwater, or terrestrial.

Nematomorpha Phylum of the hairworms, within the kingdom Animalia.

Nemertea Phylum of the proboscis worms, within the kingdom Animalia

neritic zone The open surface waters that lie over the continental shelf, from the low water level to the area where the shelf drops away into PELAGIC zones. It is the most productive part of the ocean, sheltering species such as herring, eel, mackerel, bluefish, tuna, and sharks.

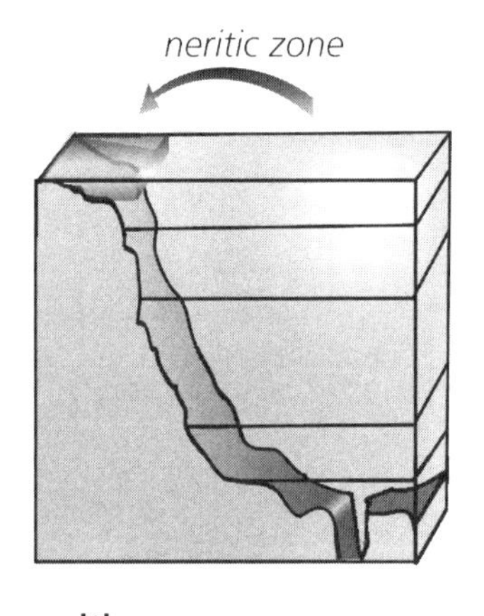

neritic zone

neuston Organism existing in the first few fractions of an inch of the water's surface. Examples are halobates, some spiders and protozoans, occasional worms, snails, insect larvae, and hydras. Neuston are distinguished from the plankton, which only incidentally becomes associated with the surface film.

Newfoundland Basin A large oceanic basin of the mid-North Atlantic centered at about 44° N and 40° W between the undersea Gibbs-Charlie Fracture Zone to the north and the Oceanographer Fracture Zone to the south. Newfoundland, Canada, lies to the west and the

Azores to the southeast. The point of greatest depth is roughly 16,600 feet (5,059 m).

New Hebrides Trench An L-shaped trench of the South Pacific's CORAL SEA. The crux is centered at about 24° S and 171° E along the TROPIC OF CAPRICORN, east of New Caledonia. Its deepest point is roughly 25,281 feet (7,705 m).

niche The role and position of an organism in its ecosystem; the full range of biological and physical conditions under which it lives and reproduces and its interactions with other species.

niskin bottle A water and plankton sampling bottle usually made of unreactive PVC (polyvinyl chloride) (to minimize possible contamination) with two trap doors, one on each end. Attached to a predetermined length of cable, the bottle is lowered, and when the desired depth is reached, a metal weight is sent down the cable; when it reaches the bottle, it snaps the doors closed, trapping the sample inside.

nitrogen narcosis A condition commonly known among scuba divers as "rapture of the deep," it is a state of euphoria that sets in when divers go below a depth of about 130 feet (40 m), or approximately seven times atmospheric pressure. Caused by the narcotic effects of the air's nitrogen at high pressure, symptoms include loss of judgment that may cause the individual to discard equipment or engage in other foolish behavior. This condition may also affect helmet suit divers below a depth of about 200 feet (61 m).

NOAA *(National Oceanic and Atmospheric Administration)* An organization of the U.S. government whose mission is to describe and predict changes in the Earth's environment and conserve and wisely manage the nation's coastal and marine resources.

See also Useful websites in the Appendixes.

Nordic Seas A term used collectively to describe the GREENLAND and NORWEGIAN SEAS.

North American Basin Centered at about 30° N and 60° W, this is a large area in the North Atlantic Ocean containing the SARGASSO SEA. The Sohm Abyssal Plain borders it to the north, the Mid-Atlantic Ridge to the east, the Guiana Basin and South America to the south, and North America to the west.

North Atlantic Current A surface current of the North Atlantic Ocean setting northeastward past the Grand Banks toward the United Kingdom to form the NORWEGIAN CURRENT. It is an extension of the Gulf Stream.

which carries to the northern regions warmer water that helps moderate the otherwise cold climate.

North Australian Basin A deep little basin of the eastern Indian Ocean centered at about 14° S and 116° E between Australia, to the south, and Indonesia, to the north. The point of greatest depth is roughly 22,587 feet (6,884 m).

North Brazil Current A current of the South Atlantic Ocean setting southwestward along the coast of Brazil. It is an extension of the westward-flowing SOUTH EQUATORIAL CURRENT as it encounters the South American coast and splits.

North Cape Current A current flowing into the Arctic Ocean. Setting eastward, it rounds the coast of Norway, carrying waters more warm and saline than those of the BARENTS SEA, into which it feeds. It is an extension of the NORWEGIAN CURRENT.

Northeast Pacific Basin A large oceanic basin of the North Pacific Ocean existing between the western coast of North America and the INTERNATIONAL DATE LINE. Undersea geographic features include the west-east-running ribs of the Mendocino, Murray, Molokai, Clarion, Clipperton, Galapagos, and Marquesas Fracture Zones.

Northeast Passage A water route of the Arctic Ocean along the northern coast of Europe and Asia sought as early as the 15th century by English, Dutch, and Russian navigators for the purpose of establishing an oceanic trade route to India and the Spice Islands. In 1878–79 the Swedish navigator NILS A. E. NORDENSKJÖLD first successfully traversed the route, which runs from Scandinavian ports, across the Arctic Ocean, to exit through the Bering Strait between Alaska and Russia.

North Equatorial Current A westward-flowing current of the North Atlantic and North Pacific Oceans lying immediately north of the equator. It also occurs in the Indian Ocean from about October to July.

Northern Hemisphere The half of the Earth north of the equator.

North Pacific Current A broad current of the mid–North Pacific Ocean branching eastward from the Kuroshio and Oyashio Extensions, where it forms part of the North Pacific subpolar gyre.

North Pole Geographically, the North Pole sits at latitude 90° North. It is the northern point of the Earth's axis and is distinguished from MAGNETIC NORTH. In 1909 the American explorer ROBERT PEARY was the first to reach the North Pole.

See also MAGNETIC DECLINATION.

North Sea An epicontinental sea of the North Atlantic Ocean centered at about 56° N and 3° E. The NORWEGIAN SEA borders it to the north, northwestern Europe to the south, Great Britain to the west, and Denmark and Norway to the east.

Northwest Pacific Basin A basin of the North Pacific Ocean centered at about 35° N and 160° E off the eastern coast of Japan. Undersea boundaries include the EMPEROR SEAMOUNTS to the east, the KURIL TRENCH to the north, and the Mid-Pacific Mountains to the south. The greatest depths are roughly 21,050 feet (6,416 m).

Northwest Passage A water route through the Arctic Ocean along the north coast of Canada that serves as a northern passage between the Pacific and Atlantic Oceans. Its discovery, continually sought since the 16th century, removed the barrier to trade between Europe and East Asia caused by the necessity to sail around North America. After many had tried and failed, in 1903–06, the Norwegian explorer ROALD AMUNDSEN was the first to navigate a path through the Canadian Archipelago and thus establish the existence of a Northwest Passage.

Norwegian Basin A basin of the Arctic Ocean centered at about 68° N along the prime meridian between Greenland and Scandinavia. The point of greatest depth is roughly 10,900 feet (3,322 m) in the Dumshaf Abyssal Plain.

Norwegian Current A current of the North Atlantic Ocean setting northeastward off the coast of Norway, where it forms the NORTH CAPE CURRENT. It is an extension of the NORTH ATLANTIC CURRENT.

Norwegian Sea One of the so-called NORDIC SEAS belonging to the Arctic Ocean. It is centered approximately at the intersection of the prime meridian (0°) and the Arctic Circle (66.30° N) between Greenland, to the west, and Scandinavia, to the east.

Norwegian Trench A trough in the seafloor lying between the southern coast of Norway and the edge of the continental shelf that stretches beneath the NORTH SEA. It is centered at about 58° N and 5° E; average depths are between 800 and 2,000 feet (244–610 m).

nudibranch A soft-bodied, gastropod mollusk known as a sea slug. It lives in shallow waters; has prominent branched outgrowths called cerata, which act as gills; and subsists largely on sea anemones. Sea slugs belong to the order Nudibranchiata, subclass Opisthobranchia, class Gastropoda, phylum Mollusca, kingdom Animalia.

North Sea – nudibranch GLOSSARY

nutrients In the ocean, any one of a number of inorganic or organic oceanic substances necessary as nutrition for primary producers, for example, nitrogen and phosphorus compounds.

ocean The largest and most uninterrupted expanses of open salt water of the Earth. Covering roughly 71 percent of the surface of our world, they are divided into five major parts, the Pacific, Atlantic, Indian, Arctic, and Southern Oceans.

ocean basin *See* ABYSSAL PLAIN.

oceanic crust The Earth's thin, dense crust beneath the ocean basins. Occasionally it may extend under a continent. The oceanic crust, the youngest rock on Earth, is largely composed of basalt.

oceanic desert A region of ocean deficient in nutrients, where plant and animal life is noticeably absent.

oceanic trench A deep, narrow cleft in the ocean floor. *See* HADAL ZONE.

oceanic zone The PELAGIC environment beyond the continental shelf break.

oceanodromous fish Fish that spend all their lives in the ocean. Compare to ANADROMOUS and CATADROMOUS FISH.

oceanographic vessels Ships used primarily for the scientific exploration and study of the seas. An account of many research vessels and their voyages can be found throughout this book.

oceanography The study of the oceans and the marine environment involving such specializations as biology, botany, zoology, geology, geography, and archeology.

oceanology The study of the science of the oceans.

ocean ridge An underwater, midocean mountain range where volcanic activity produces the new material for SEAFLOOR SPREADING and therefore continental drift.

ocean troposphere A term sometimes used to describe a zone of relatively high temperature and strong currents occupying the upper layer of the ocean.

octant An early instrument of navigation designed to measure the altitudes of celestial bodies and therefore work out a ship's latitudinal position. It is of the same principle as the SEXTANT but employs a 45° angle, as opposed to the sextant's 60° or the QUADRANT's 90° angle.

octopus An eight-armed marine invertebrate MOLLUSK, carnivorous, that can change colors rapidly to blend with its environment. It averages about three feet (91 cm) in length but the little pygmy octopus of the Atlantic reaches only about five inches (13 cm). Octopuses are deaf (as are other cephalopods) but have acute vision, are found worldwide, and spend most of their time hiding in rock crevices, luring victims close by wiggling a tentacle to imitate a worm. It moves by jet propelling water through its body or pulling itself along with its arms. Octopuses belong to the genus *Octopus*, family Octopodidae, order Octopoda, class Cephalopoda, phylum Mollusca, kingdom Animalia.

See also COMPLEX CAMERA EYE.

Okhotsk, Sea of Centered at about 55° N and 150° E, this is a shallow, marginal sea of the North Pacific Ocean indenting eastern Siberia.

oligotrophic Descriptive of waters that contain low concentrations of nutrients. *See also* EUTROPHIC; MESOTROPHIC.

Oman, Gulf of A body of water centered along the TROPIC OF CANCER at about 59° E. It connects the ARABIAN SEA, to the east, with the Persian Gulf, to the west. To the south and north it is flanked by the Sultanate of Oman (southern Arabia) and southern Iran, respectively.

omnivore A term applied to animals that eat both plant and animal matter. The Japanese shore crab is an example of a marine omnivore, existing on clams, scallops, oysters, and algae.

ontogeny Pertaining to biology, it is the development of a species from origin (fertilization) to adult form.

ooze A soft MARINE SEDIMENT consisting of animal skeletal remains and inorganic materials. An ooze can be classified by the contributing organisms of the region.

See also COCCOLITHS; DIATOM; GLOBIGERINA OOZE.

orca A marine mammal also known as the killer whale, carnivorous, and the largest member of the dolphin family. The males grow to be twice as large as females, around 30 feet (9 m) long. They are black with a bright white patch around each eye and on their bellies, which creeps up their sides. Found worldwide, they travel in family packs feeding on seals, birds, and fish. Once, in Wrangell Narrows, Alaska, a resident witnessed and reported a pod of killer whales that was organizing an ambush on a pod of porpoises, letting the young orcas do the slaughtering while the older ones blocked all escape. Reportedly, the channel was bathed in red from the frenzy. Orcas belong in the family Delphinidae, order Cetacea, class Mammalia, subphylum Vertebrata, phylum Chordata, kingdom Animalia.

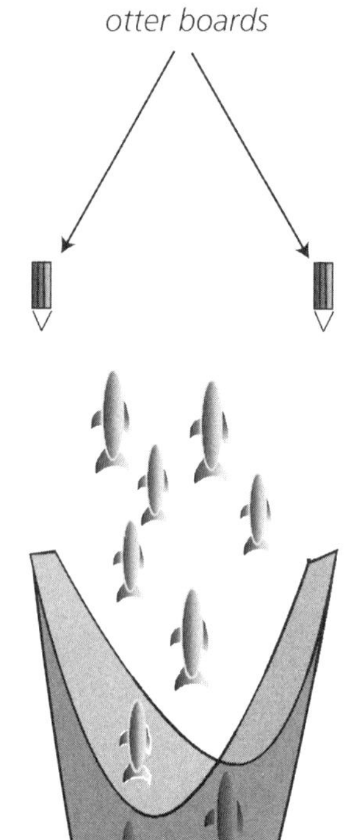

otter trawl

order In TAXONOMY, a classification group consisting of closely related genera, below a class and above a family.
See also KINGDOM.

organic Referring to anything derived from, related to, or having properties associated with living things or once-living things.

Orthonectida and Rhombozoa Phyla of the mesozoans, or parasitic flatworms of marine invertebrates. These fall within the kingdom Animalia.

osmoregulation In an organism, the physiological process by which it regulates the salinity of fluid in its body to tolerable levels.

osmotrophic Used to describe a heterotrophic species that obtains nutrients by absorbing organic matter that is in solution from its surroundings. *See also* PHAGOTROPHIC.

otter trawl A fishing net towed behind a vessel that is used to collect pelagic organisms both on and near the bottom of the ocean. Unlike the beam trawl, which keeps the net constantly open with a square frame, the otter trawl uses boards called otter boards that attach to each side of the net and are forced apart by water pressure. When the vessel slows to a stop, the trawl net closes automatically.

overfalls Breaking waves caused by two currents that meet or by waves that are traveling against the current.
See also TIDE RIP.

overfishing The rate at which the harvesting of a species of fish exceeds the rate of reproduction. *See* MAXIMUM SUSTAINED YIELD.

oviparous A term relating to a female animal that lays eggs of young that hatch outside the body.

ovoviviparous A term relating to a female animal whose eggs hatch inside the body and whose young are born later.
See also VIVIPAROUS.

Oyashio Current A current of the North Pacific Ocean setting southwestward from the BERING SEA along the eastern coast of Siberia, Kamchatka, and the Kuril Islands until it merges with the northerly flowing Kuroshio Current and sets out to sea.

oxygen cycle The circulation and reprocessing of oxygen in the biosphere.

oyster A bivalve marine mollusk found worldwide. The oyster is a sessile animal, liking shallow waters, that in its larval stage is free-swimming. Most types are edible and have been harvested for ages, since early Roman times. The American oysters grow from two to six inches (5–15 cm) in length and are continually preyed upon by starfish and humans alike. Oysters are easily farmed, and from the great pearl oyster, which can reach 12 inches (30.5 cm) in the tropical species, we harvest pearls. Oysters belong in the family Ostreidae, order Filibranchia, class Pelecypoda or Bivalvia, phylum Mollusca, kingdom Animalia.

oyster reef A thick bed of oysters found in estuarial environments.

pacific deep water A cold-water mass circulating slowly in the bottom of the Pacific Ocean, ranging from 3,300 to 10,000 feet (1,006–3,048 m) in depth, originating in the Southern Ocean and spreading north into the Pacific.

Pacific Ocean Earth's largest ocean of the five divisions of the WORLD OCEAN. Its surface area covers about 60,061,000 square miles (155,557,000 km^2), extending from the Arctic to the Southern Ocean, and from Asia and Australia in the east to the Americas in the west. *See also* ARCTIC OCEAN; ATLANTIC OCEAN; INDIAN OCEAN; SOUTHERN OCEAN.

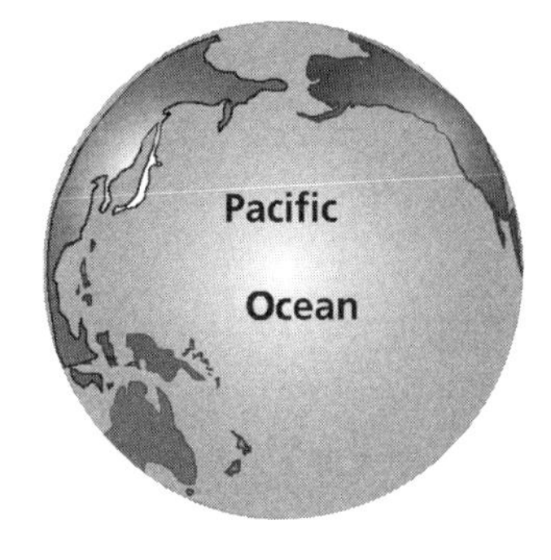

Pacific Ocean

Pacific-type margin The leading edge of a continent, as opposed to an ATLANTIC-TYPE MARGIN, which is the trailing edge of a continent. *See* CONTINENTAL DRIFT.

pack ice Floating ice masses that have packed together, making a sea impossible to navigate without a stout ship designed as an icebreaker. Influenced by wind above and current below, the pressure at times is so fantastic that often the white floes ramp up on each other to create a surreal landscape of slanting, castlelike pinnacles seen nowhere else on Earth. Pack ice can be found in both the Arctic and Antarctic Oceans. Creatures such as penguins, seals, and even polar bears make pack ice their home above water.

painter line As a nautical term, used to describe the line attached to the bow of a boat being towed.

palaeoceanography A branch of oceanography that examines the biological and physical characteristics of the oceans during ancient times.

pallium The MANTLE of a MOLLUSK or brachiopod.

Pangaea Prehistoric supercontinent of the young Earth from which all other continents have formed; over the eons, it has played the role of stoic witness to the millions of changes in our evolution.

See also CONTINENTAL DRIFT; GONDWANALAND; LAURASIA; PLATE TECTONICS.

Panthalassa Name given to the early world ocean surrounding Pangaea. Ocean circulation was simple then, when the waters reached from pole to pole with only a single arm for the TETHYS SEA. Because of slow mixing, the temperature gradient would have been low, exhibiting little difference between the equator and the poles.

paralytic shellfish poisoning (PSP) A poisoning that is caused by neurologically damaging saxitoxins that are by-products of a form of microscopic algae (dinoflagellates). People who eat shellfish that have been feeding on toxic dinoflagellates can suffer numbness, paralysis, disorientation, and even death.

parapodia Paired, lateral extensions on each segment of a POLYCHAETE (worm) that may be used in swimming, burrowing, or crawling.

parasite An organism living on or in the body of another organism (the host), deriving nourishment at the expense of the host, usually without killing it.

parasitism An association between different organisms whereby one benefits and the other is harmed. Every living species is potentially subject to parasitism.

parts per billion (ppb) The number of parts of a chemical found in one billion parts of a particular liquid, gas, or solid mixture.

parts per million (ppm) The number of parts of a chemical found in one million parts of a particular liquid, gas, or solid mixture.

parts per trillion (ppt) The number of parts of a chemical found in one trillion parts of a particular liquid, gas, or solid mixture.

patchiness Pertaining to PLANKTON, the uneven distribution of organisms.

pearl A bead that forms around a foreign particle, typically a grain of sand, that finds its way inside a MOLLUSK.

pecten *See* SCALLOP.

pectoral fin On a fish, the fin or fins along the side of the body.

pedal locomotion A type of locomotion whereby an organism's muscular system flexes with waves of activity, and that flexing in turn propels it across the seafloor. Flatworms, some cnidarians, and gastropod mollusks exhibit this means of self-transport.

pelagic Pertains to the open ocean, away from coasts and continental shelves.

pen In squid, a thin, chitinous internal shell. *See* CUTTLEFISH.

penguin A flightless bird of Antarctica that swims with flipperlike wings and steers by using webbed feet as rudders. They can reach speeds of up to 25 miles per hour (40.2 kph) as they pursue fish, squid, and shrimp. Their stiff, waterproof feathers serve as insulation. Most species are black with a white breast. The largest are the emperor and king penguins, which reach three to four feet (1–1.2 m) in height. Penguins belong in the family Spheniscidae, order Sphenisciformes, class Aves, subphylum Vertebrata, phylum Chordata, kingdom Animalia.

peninsula A land track extending out into a sea or lake. The state of Florida is called the Floridian Peninsula.

pennant line Also called pendant, a nautical term describing the line by which a ship is made fast to a mooring buoy.

pentaradial symmetry *See* ECHINODERM.

pereiopod Thoracic appendage of a CRUSTACEAN.

perennibranchiate Bearing permanent gills, especially certain salamanders.

periostracum The horny outer layer of a MOLLUSK shell.

permeability The condition in which a substance or membrane allows the passage of liquid.

Persian Gulf Centered at about 25° N and 53° E, this is a body of water connected to the ARABIAN SEA by way of the Gulf of Oman. It separates Iran, to the north, from Saudi Arabia, to the south, with tiny Kuwait between them on the west.

Peru Basin A basin of the South Pacific Ocean centered at about 18° S and 90° W off the coast of Peru, South America. Its point of greatest depth is roughly 16,630 feet (5,069 m) near the Sala-y-Gomez Ridge.

Peru-Chile Trench A seismically active trough created by subduction and SEAFLOOR SPREADING. Extending north-south from longitudes 10° S through 60° S, it runs nearly the entire length of South America's western coastline amid the subducting Nazca Plate on the west and the South American Plate on the east. The deepest point is roughly 25,000 feet (7,620 m).

Peru Current Originating in the WEST WIND DRIFT in Antarctic waters, it is a cold, broad surface current of the Pacific Ocean setting northward along the midwestern coast of South America. It is also known as the Humboldt (PERU) Current, named for the German scientist Alexander von Humboldt, who first charted its flow in 1802.

petaloids The petal-like appearance of respiratory podia in irregular sea urchins and the formations of petaloid Scleractinian corals such as Well's petaloid coral (*Coscinaraea wellsi*) and the green petaloid coral (*Psammocora explanulata*).

Peterson grab Clamshell-like instrument made of zinc-plated steel used for bottom sampling in search of benthic organisms.

Phaeophyta Phylum of the brown algae, within the kingdom Plantae.

phagotrophic Used to describe a species of heterotrophic PHYTOPLANKTON that feeds on other phytoplankton or DETRITUS.
See also OSMOTROPHIC.

pheromone Chemical hormone released as a signal by one organism that influences the behavior of another organism of the same species, used as territorial markers, food locators, or attractants for the opposite sex. Female crabs, for example, employ this hormone to attract males.

Philippine Sea A sea of the North Pacific Ocean centered at about 18° N and 135° E. The Philippine Islands border it to the west, the Mariana Islands and the Pacific Ocean to the east, Japan to the north, and the lacework of the Caroline Islands to the south.

Phleger bottom sampler An ocean bottom core sampler, designed to take core samples of upper sediment layers, with which one might study an area's quantity of FORAMINIFERA. It is a short, weighted tube with a removable lining and a cutting edge. Using a winch, it can be operated as deep as 15,000 feet (4,572 m) if the sediment is a sticky type.

phonation The production of sound.

phoresy A passive form of transportation, especially in some arthropods and fishes, in which they cling to an animal of a different species and then release themselves after traveling for some distance. *See* REMORA.

Phoronida Phylum of the sessile, chitinous, tube-dwelling wormlike marine invertebrates, within the kingdom Animalia.

phosphorus An essential element in living organisms, important in providing energy for muscle contraction in vertebrates (creatine phosphate) and invertebrates (arginine phosphate). It is present in seawater as inorganic ions, as phosphorite on continental shelves, and as soluble deposits of complex salts. PHYTOPLANKTON deplete water supplies of

GLOSSARY petaloids – phosphorus

phosphorus, using it to produce phosphoproteins, phospholipids, and DNA (deoxyribonucleic acid).

photic Of or relating to light.

photoautotroph An organism that synthesizes organic matter by using the energy of light.

photophore A complex organ that produces bioluminescent light in fish.

photosynthesis The process employed by plants through which carbon dioxide and water are used to make energy-rich organic compounds when CHLOROPHYLL and sunlight are present. Most forms of photosynthesis release oxygen as a by-product. Scientists estimate that around 90 percent of the world's oxygen is created by photosynthetic processes derived from within the ocean.
See also ALGAE.

phototaxis The stimulated movement of an organism in response to a light source. If moving away from the light, it is called negative phototaxis. If moving toward the light, it is called positive phototaxis.

phycology Also known as algology, the scientific study of algae and particularly seaweeds.

phylogeny The relationship among species through a succession of forms as reflected by their evolutionary history, beginning with the ancestral form and then branching off into descendants.

phylum In TAXONOMY, a principal division of all plants and animals from which point subdivisions begin. The only division higher than a phylum is a kingdom; the next division below a phylum is a class.
See KINGDOM.

phytophagous Descriptive of organisms that feed on plants.

phytoplankton Microscopic, usually one-celled plant plankton (as opposed to zooplankton, which is animal plankton), such as DIATOMS and DINOFLAGELLATES.

picoplankton Extremely small plankton, mostly bacterioplankton, with cells less than 0.2–2 µm in size, too small to catch in a standard plankton net.

pile Also called piling, a heavy wood, metal, or concrete pole driven into the sea or lake bottom to support piers, wharves, or docks.

pinniped Named from Latin words meaning "fin-footed," marine mammals characterized by having four swimming flippers, such as seals and sea lions.

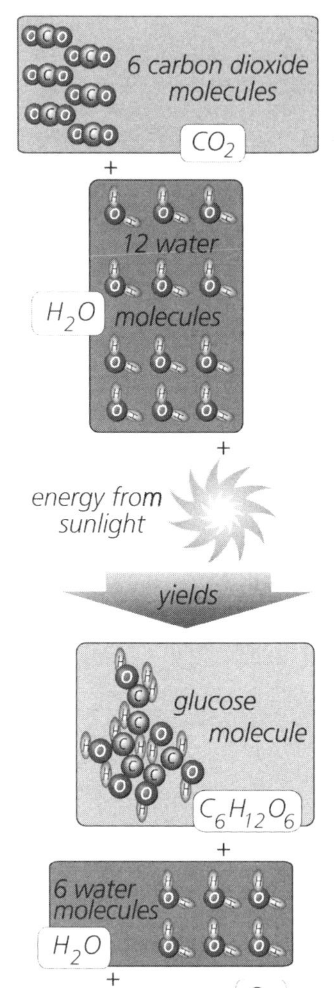

photosynthesis

piscivores Fish-eating animals.

pitchpoling A nautical term meaning a circumstance in rough, heavy seas in which a ship is being thrown end over end.

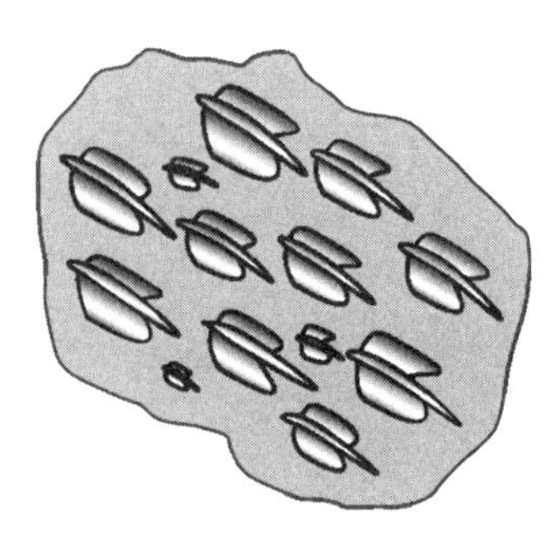

placoid scales

placoid scales Also sometimes known as denticles, these are hard SCALES, small and toothlike in structure, covering the bodies of cartilaginous fishes such as sharks and rays. The skins of these animals were used commercially for hundreds of years to make nonslip grips for the handles of swords and knives.

Placozoa Phylum of the simple metazoans, within the kingdom Animalia.

plankton Passively drifting or weakly swimming organisms suspended in the WATER COLUMN, relying on ocean currents for transport and distribution.

plankton bloom Evident as green, turbid waters, this is a high regional concentration of plankton resulting from an extreme level of plant materials in the presence of plentiful sunlight and nutrients.

Plantae The plant kingdom. Within this section, many marine-related phyla are recognized.

plate tectonics A theory first developed in 1912 from the reasoning of the German scientist ALFRED WEGENER and better formulated in the 1960s, that the Earth's crust is made up of thin, rigid lithospheric plates atop a more liquid asthenosphere, which move in relation to one another. Though the idea was not well received in the beginning, it is widely accepted today. It offers an obvious scientific explanation for the way the continents seem to fit as puzzle pieces and the way they apparently drifted apart; the placement of the mountain ranges, SEAFLOOR SPREADING, the reasons volcanoes erupt and some exist along a belt; and the presence of magnetic anomalies on the seafloor. Some plates are moving away from each other and some toward, at an average rate of two to three inches (5 to 7.6 cm) per year but sometimes race at 7.3 inches (18.5 cm) per year.

See also CONTINENTAL DRIFT.

Platyhelminthes Phylum of the flatworms (platyhelminths), flukes, and tapeworms, within the kingdom Animalia.

pleuston Pertains to animals that float passively on the surface of the ocean projecting parts of their body into the air as sails, letting the wind assist in propelling them along with the current.

pneumatocyst *See* SWIM BLADDER.

pod A group of whales.

Pognophora Phylum of the giant deep-water tube-dwelling worms, within the kingdom Animalia.

poikilotherms Cold-blooded animals. *See* HOMOIOTHERMS.

point On a compass, one of 32 points equal to 11.25 degrees.

polar Pertaining to the cold regions of the Earth or any other celestial body. On Earth they exist above the Arctic Circle (latitude 66.30° N) in the Northern Hemisphere and the Antarctic Circle (latitude 66.30° S) in the Southern Hemisphere, experiencing little precipitation all year.

polar biome Term used when describing the flora and fauna of the northern and southern polar regions of Earth. The Arctic North biome enjoys more variety of life and depends more on the PHYTOPLANKTON and ZOOPLANKTON food chain than does the biome of the Antarctic South.

Polaris One of many names given to the North Star. From Earth's vantage, it is a relatively fixed point in the visible night sky used for centuries as a celestial navigation bearing.

poles The precise axis points of a revolving celestial body. On Earth, the North and South Poles.

pollution Any material introduced into the environment that is considered harmful to the ecosystem. Pollution takes many forms: water, air, soil, solid waste, hazardous waste, and even noise. Some pollution is environmental, such as that resulting from volcanic eruption, but most is derived through human activity. However, unlike any other species on the planet, we have the technology, the intellect, the will, and the desire to control it.

polychaetes (Courtesy of NOAA)

polychaetes Abundant benthic marine segmented worms, some in tubes, some free-swimming, found from the neritic to the hadopelagic zones. Polychaetes make up the class Polychaeta and compose the majority of the phylum Annelida within the kingdom Animalia.

polychromatic Many-colored. Brilliantly colored tropical fish are polychromatic.

polyculture Cultivation of more than one species of organism in an aquaculture system.

polygynous In male animals, mating with more than one female in a season.

polyp An individual cnidarian (COELENTERATE), cylindrical in form, with a mouth at one end surrounded with tentacles. It attaches itself to the substratum and forms part of a colony, such as the reef-building coral.

Porifera Phylum of the sponges, within the kingdom Animalia.

porpoise A fun-loving marine cetacean related to the dolphin. The porpoise and dolphin are often confused with one another. A porpoise is smaller than most dolphins, reaching about seven feet (2 m) in length, and does not possess the dolphin's beaky snout, having a more rounded nose. Their colors are typically dark blue to black with white underneath. They exist in cool to cold protected waters of the Northern Hemisphere and sometimes play in the bow wake of a vessel while it is under way, making the characteristic explosive exhale as their backs hump above the surface for a breath. They mainly eat squid but also eat fish and crustaceans. Porpoises belong in the family Phocoenidae, order Cetacea, class Mammalia, subphylum Vertebrata, phylum Chordata, kingdom Animalia.

port 1) A shipping harbor. 2) On a vessel, when facing forward it is the left side of the ship. Port is indicated by a red light aimed toward the bow, visible to all traffic but that which travels astern. Compare to STARBOARD.

Portuguese man-of-war Free-swimming marine cnidarian (COELENTERATE) with long stinging tentacles and brightly colored inflated bladders with a protruding crest, which catches the force of the current and helps casually propel them along. These tropical colonial animals reach about six inches (15 cm) in length, but their ribbons of tentacles can sometimes trail to 100 feet (30 m) and are armed with a powerful neurotoxin. The stinging cells (nematocysts) remain active even after the animal is dead. The Portuguese man-of-war belongs to the order Siphonophora, class Hydrozoa, phylum Cnidaria, kingdom Animalia.

posterior The tail end of an animal. *See also* ANTERIOR.

potential evaporation The total amount of water that could possibly evaporate from an unlimited water supply, as opposed to actual evaporation, which is the amount evaporated from a somehow limited water supply.

prawn A name for a commercially valuable marine crustacean, of which there are many species. Prawns have 10 legs, grow to about four inches (10 cm) in the European variety, and differ from SHRIMP in having a long, serrated beak (rostrum); long antennae; slender legs; and laterally compressed abdomens. They live along sandy shores in freshwater and brackish water, in tropical and temperate zones.

precession of the equinoxes A slow wobble in the axis of the Earth's rotation that causes a gradual change in the direction in which the axis points. This has many effects on the Earth, one of which is a variation in the tidal ranges over time.

precious corals GORGONIAN CORALS that secrete a red or pink skeleton.

predation The stalking, capture, and ingestion of prey by a predator for energy and nutrients in order to maintain its life.

pressure Simply put, the application of force on the surface of an object. The pressure at sea level is measured as one atmosphere of pressure, or 14.7 psi (pounds per square inch) (1.01 bar). At 18,000 feet of altitude (5,486 m) there is only one-half the amount of pressure at sea level. As you sink beneath the waves, for every 33 feet (10 m) you descend you gain one atmosphere of pressure, so at a depth of 66 feet (20 m) you are subject to the pressure of three atmospheres, or 44.1 psi (3.04 bar). *See* BAR.

Priapulida Phylum of the marine cucumber and wormlike burrowing invertebrates, within the kingdom Animalia.

primary production Amount of organic material produced by organisms from inorganic material in a particular volume of the WATER COLUMN within a certain length of time. Most primary productivity in the ocean is due to PHOTOSYNTHESIS of phytoplanktonic algae.

prime meridian More commonly called the Greenwich meridian, it is an imaginary line of longitude passing through Greenwich, a borough of London, England, that serves as the reference point from which all other lines of longitude are measured, east or west, on the world globe. In 1884, during a meeting held in Washington, D.C., by an international conference of astronomers, the decision was made that the line of longitude passing through Great Britain's Royal Observatory should become the meridian of 0° degree. This system is used to divide the world into 23 full time zones (15° wide) and two half-zones (7.5° wide). This knowledge is essential when using a chart to plot a course over open waters.

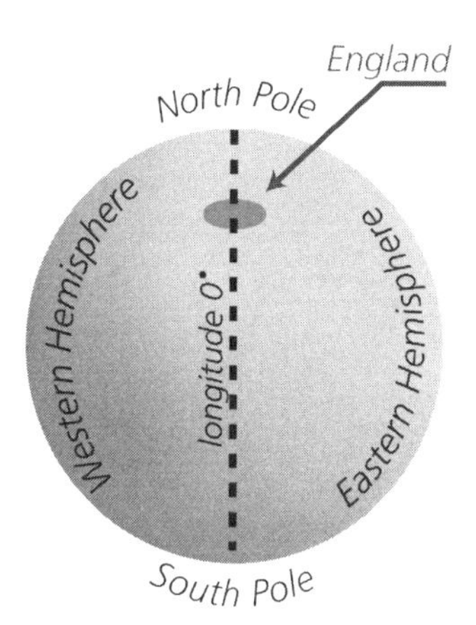

prime meridian

proboscis A tubular protuberance on the head of an animals, such as a marine pinniped. It is descriptive of an enlarged nose, such as that found on the elephant seal.

productivity Amount of organic material produced by organisms in a particular volume of the WATER COLUMN within a certain length of time.

promontory A high, pointy landmass extending into the ocean.

proprioceptor A sensory organ located deep within body tissue, especially muscle, which is receptive to stimuli from within the body, such as stretching, movement, and body position.

Protista Also called protists, in some classifications this constitutes a kingdom of simple organisms that include the PROTOZOANS, fungi, bacteria, and algae. It was first proposed in 1866 by the German zoologist ERNST H. HAECKEL and was eventually introduced to help overcome difficulty in classifying organisms that display both animallike and plantlike characteristics. Most protists reproduce through cell division, though some reproduce sexually.

protochordates Chordates that lack a backbone, such as sea potatoes, sea peaches, and sea squirts.

See also TUNICATE.

protoplasm Pertaining to the living matter of cells; its use is not commonplace among today's professionals because of its lack of specificity. The term was introduced by the Czechoslovakian physiologist Johannes Purkinje in 1839, when scientists of the time knew that all living things were made from cells of jellylike substance, termed protoplasm.

Protozoa The classic kingdom of single-celled, animallike organisms, some of which may form colonies. Protozoa is Latin for "first animals"; the Protozoa are a very diverse group that can exist on little or no oxygen and range from plantlike forms to groups that feed and behave as animals do. Protozoans can be found everywhere: in freshwater and salt water, air, and damp earth, and even within other animals and plants. Within this section, many phyla of the kingdom Protozoa are recognized.

See also PROTISTA.

protozoan Unicellular eukaryotic (having a true nucleus) heterotroph (an organism that eats other organisms).

protozoology The scientific study of protozoans.

pseudopod A term meaning "false foot," which refers to a projection that serves as a foot or means of locomotion, found in organisms such as some amoeba, protists, protozoans, and foraminifera.

psychrosphere One of two zones (the other is the THERMOSPHERE) into which the depths of the sea are sometimes separated according to temperature. The psychrosphere are waters with temperatures less than 50° F (10° C) and usually fall in a range anywhere between

328 and 2,300 feet (100–700 m) in depth, depending on the area. This zone coincides with the OCEAN TROPOSPHERE.

pteropod Pelagic marine gastropod (snail) found in surface layers of the ocean whose foot is modified for swimming and shell is reduced or absent. Examples are the shelled Limacina helicina, an Antarctic pteropod that lives on phytoplankton, and the shell-less sea butterfly, which is carnivorous.

Puerto Rico Trench An oceanic trough lying north of the island of Puerto Rico that at the time of this writing contains the deepest known point in the Atlantic Ocean, 28,000 feet (8,534 m), called the Milwaukee Deep. The trench is centered at about 19° N and 65° W.

purse seine A type of net used by commercial fishers to harvest pelagic fish. Two boats are needed to operate the net, the seine boat and the seine skiff. The skiff is launched from the main vessel (from which the net is permanently established) and tows the free end, pulling the net closed, until is resembles a purse, around schools of fish. Different meshed nets can be used, depending on the species desired in the harvest.

pycnocline In the WATER COLUMN, a transitional layer between the HALOCLINE and the THERMOCLINE, constituting about 18 percent of the ocean basin, in which the water density changes with depth. In lower latitudes, the pycnocline coincides with the thermocline, and in midlatitudes, with the halocline.

Pyrrhophyta Phylum of the dinoflagellates, within the kingdom Plantae.

quadrant Precursor of the modern SEXTANT, it is an angle-measuring device dating back to medieval times used to measure the altitude of a celestial body; it is based on a scale of 90°, which is distinguished from the sextant's 60° scale and the OCTANT's 45° scale. Large wall-mounted units were used from Ptolemaic times through the 18th century to take precise astronomical measurements. In about 1750–70, the London instrument maker John Bird improved the quadrant to its most accurate capabilities.

quadruped Descriptive of a four-footed animal.

quahog A hard CLAM with a thick, heavy shell in a vague shape of a heart that lives so shallowly in INFRALITTORAL ZONES it can be harvested with a rake at low tide; it was named such by Native Americans. Quahogs can be found in coastal waters from the Gulf of Saint Lawrence in eastern Canada to the Gulf of Mexico.

quarter The sides of a ship aft (toward the rear) of amidships.

queen conch The largest of the conches, it is a marine gastropod MOLLUSK that lives in warm waters. Its shell of overlapping whorls and outer triangular whorl can reach lengths of up to three feet (91 cm). A conch has a clawlike structure in the shell opening that it uses to propel itself along in short, sudden spurts and to dig into sediment. They are harvested for their edible meat, and the shell has been used in Europe to carve cameos. They are also sometimes used as a trumpet for ceremonial rituals. True conchs make up the family Strombidae, order Mesogastropoda, subclass Prosobranchia, class Gastropoda, phylum Mollusca, kingdom Animalia.

quicksand A deep, swampy load of very fine sand found along seacoasts and riverbeds that is very hard to distinguish from weight-bearing sand. Quicksand is terrain that, because of fluids that prevent the particles from settling, has lost its quality of firmness. Thus it is no longer supportive. It is dangerous to human and beast alike, for, once one is trapped in quicksand, any panicked attempt to free oneself results in more firm entrapment or, if very deep, sinking below the surface. Calm attempts to spread all limbs outward and float atop the sand as if it were water, then rolling gently away, is the best way to cope with quicksand.

radar *(Radio Direction and Ranging)* An electronic instrument that emits radio waves through the atmosphere that bounce off objects and return to be received by the source. The data are used to determine the location and shapes of objects that might normally be out of sight, such as in heavy rain, snow, fog, or conditions of darkness. Radar is in substantial use by the world's commercial, pleasure, and military shipping.

radiolarian A protist (unicellular) marine organism in the same class as the amoeba and foraminifera, which is distinguished from amoeba by the secretion of intricate exoskeletons called tests, which can sometimes be several tenths of an inch wide. The spherically symmetrical test is made of silica, often sporting spines, and perforated for the extension of pseudopods for capturing food. These animals are found in the euphotic zones of all oceans. Radiolarians belong to the phylum Actinopoda, kingdom Protozoa.

radiolarian ooze A deep ocean SEDIMENT composed of at least 30 percent of the siliceous remains of radiolarians. This is a deeper ooze than that formed of calcareous skeletons, for it does not dissolve as readily with depth. Diatom ooze is another type of deep siliceous ooze.

raised beach A beach that was once an active tidal zone but has risen above normal sea level because of either the land's rising or the sea's falling. *See also* RELICT BEACH; SUBMERGED BEACH.

ray A marine cartilaginous fish related to the shark. Rays are divided into seven families of fish that glide through the water with their winglike bodies, which are essentially just huge pectoral fins. Existing worldwide, they prey on MOLLUSKS and CRUSTACEANS, crushing them with blunt teeth, and range in size from a few inches to over 20 feet (6.1 m) in width. The largest rays are the sea bats, or mantas, and the devilfishes, which may weigh in at up to 3,000 pounds (1,361 kg). The diamond-shaped stingray has a long, whiplike tail equipped with one or more large, sharp poisonous spines, sometimes barbed, that is capable of inflicting critical wounds. Ray families are the skates (family Rajidae), devilfish (family Mobulidae), electric rays (family Torpedinidae), stingrays (family Dasyatidae), eagle and bat rays (family Myliobatidae), guitarfish (family Rhinobatidae), and sawfish (family Pristidae), all belonging to the order Batoidea, subclass Elasmobranchii, class Chondrichthyes, subphylum Vertebrata, phylum Chordata, kingdom Animalia.

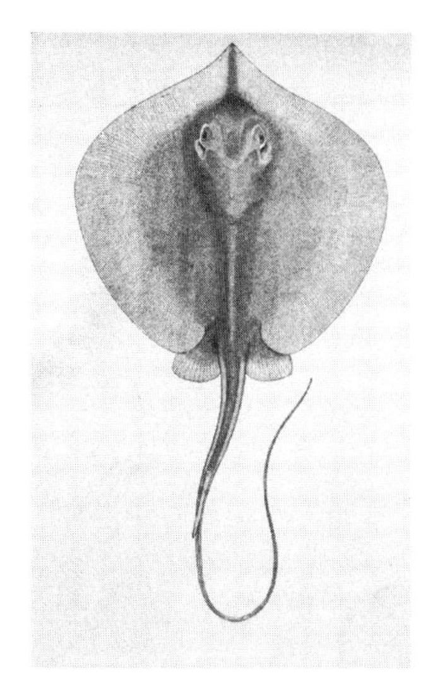

ray (Courtesy of NOAA)

razor clam A marine BIVALVE MOLLUSK distinguished by its shape, which resembles that of an old-fashioned razor. They are edible and prized by humans for their flavor. The Atlantic razor clam is a quick burrower, swims with its foot, and can grow as large as 10 inches (25 cm) long.
See also CLAM.

red clay A deep ocean SEDIMENT composed mainly of particles of continental earth borne to the oceans by wind. The southwestern Pacific Ocean is dominated by red clay sediment.

Red Sea Centered at about 21° N and 38° E, this is a long narrow body of water connected to the ARABIAN SEA by way of the Gulf of Aden. Africa borders it to the west, the Arabian Peninsula to the east, Sinai to the north; the Straits of Bab al Mandab lead to the open waters of the south.
See also SUEZ, GULF OF.

red tide The common term used for an algal bloom with an unusually high number of microorganisms that appear on the surface of the sea when currents meet to form a downwelling. Carrying nutrient-rich water, the organisms swim toward the surface as other water flows downward. They appear as long streaks in the water in varying shades of red or brown. This event can sometimes be disastrous to sea life by exhausting all nutrients from the area, limiting sunlight, causing temperature changes, or emitting poisonous toxins as a by-product. In 1977, a huge red tide off New Jersey caused the destruction of countless organisms, mainly those residing on the seabed such as clams and mussels. Red tides are also responsible for PSP, or PARALYTIC SHELLFISH POISONING, a worldwide phenomenon.

reef 1) A track of hard rock or coral in shallow seas usually submerged at all times by a thin layer of water. 2) A nautical term meaning to reduce the sail area.
See also CORAL REEF.

reef slope On a REEF, it is a steep side that sweeps downward from the edge of the reef break to a point where it begins to level out.

reef terrace It is a flat reef, like a shelf or terrace, extending seaward from land. Typically, a reef terrace is exposed at low tide.

reflection In oceanography, a phenomenon whereby surface wave energy is sent back in the direction from which it came, as when a wave connects with a cliff face and is sent offshore again.

refraction In oceanography, a phenomenon whereby the direction and speed of a surface wave are changed by physical interference.

relict beach A track of land that was once a beach but because of changing sea levels has become either submerged or left high and dry. In some submerged areas along the coast of Alaska, researchers have found Stone Age artifacts with just a few hauls of the dredge. To them, this proves that as recently as a few thousand years ago, the sea levels were much lower.
See also RAISED BEACH; SUBMERGED BEACH.

remora Marine animals, known as the sucking fishes, that spend a great portion of their lives attached to larger fish and living off the food scraps that come their way whenever the host feeds. We associate the remora with the shark, because of the shark's popularity with media, but they also attach to swordfishes, marlins, and sea turtles, and though they prefer not to, they are very capable swimmers. Occasionally they have been known to attach themselves to ships, causing one to wonder how long they wait in hunger before deciding a ship never eats. The remora varies in length, from about seven to 35 inches (18—89 cm), and its sucker apparatus, a modified dorsal fin, is located just behind the head. Remoras belong to the family Echeneidae, order Echeniformes, class Osteichthyes, subphylum Vertebrata, phylum Chordata, kingdom Animalia.
See also COMMENSALISM; PHORESY.

remote sensing The process of using specialized equipment to gather information from a distance, such as mapping the seafloor with acoustics.

reproduction The generating of all types of organisms through asexual or sexual modes that result in a type of offspring. Viruses do not reproduce but are replicated through host organisms.

respiration Metabolic process on which all living cells depend. Living things take in oxygen, use it in energy-producing chemical reactions, and then release it as carbon dioxide. It is part of the CARBON CYCLE system.

resurgence A phenomenon that occurs after a storm, in which the water level rises and falls, again and again, until the energy from oscillating waves finally expires.

Rhizopoda Phylum of the mobile amoebas, within the kingdom Protozoa.

Rhodophyta Phylum of the red algae, within the kingdom Plantae.

Rhombozoa and Orthonectida Phylum of the mesozoans, or parasitic flatworms of marine invertebrates, within the kingdom Animalia.

ría A funnel-shaped coastal inlet, typically quite deep. Many examples exist along the northwestern coast of Spain, such as the Ría de Vigo, Ría de Muros e Noya, and Ría de Pontevedra.

ribs On GASTROPOD shells, the lines that run lengthwise, and on BIVALVES, the lines that radiate.

rift valley A steep, flat-bottomed depression in the Earth formed after the collapse of the oceanic crust. Rift valleys are common to the midoceanic ridge, where the land has split and moved away as a result of pressure caused by magma that is surging upward.
See also SEAFLOOR SPREADING.

Riga, Gulf of A body of water belonging to the Baltic Sea centered at about 58° N and 23° E. It is bounded mainly by the Republic of Latvia along the south and east and the republic of Estonia to the north and connects to the BALTIC SEA to the west by way of the Straits of Irves Saurums.

rigging A nautical term used generally to describe all the lines (ropes) of a vessel. Standing rigging, for example, are the permanent lines that support the masts.

ring of fire A band of still-active undersea volcanoes, trenches, and faults that fringe most of the Pacific Ocean perimeter. Formed after the breakup of PANGAEA, it backs against the continental shelves of eastern Asia, western North and South America, and the Aleutian chain.

Rio Grande Rise A mountainous undersea ridge in the South Atlantic Ocean centered at about 31° S and 35° W. To the east lies the undersea Mid-Atlantic Ridge, to the west South America, and to the north and south the undersea BRAZIL and ARGENTINE BASINS. Its highest peak reaches to within about 2,100 feet (640 m) of the surface.

rip *See* TIDE RIP.

rip current A ribbon of intense current returning landward from the sea through the surf zone. Waves accumulate against the coast and are thus drained back to seaward by the rip current. These usually occur along jetties, forelands, and other erratic beach points, and also at regular intervals along straight, uninterrupted beaches.

ripple marks The wave patterns seen across fine, sandy ocean bottoms, caused by the action of bottom current. In an eerie way they resemble the crest and trough pattern of surface waves. Sometimes the sediment is found fossilized, preserving ancient ripple marks from times when many different and exotic species once cruised the seas.

rise In terms of the ocean floor, it is a long, broad, submerged elevation of land with a smooth and gentle rise.

roaring forties A term from the old sailing days applied to the westerly winds that blow unobstructed by any kind of large landmass through 40° S and 50° S latitude.

rode A nautical term used to describe the line or chain attached to a ship's anchor, also called the anchor rode.

rooster tailing The arcing tail of spray present at times when cetaceans such as porpoises and dolphins are swimming at high speeds.

Ross Sea Centered at about 75° S along the IDL (INTERNATIONAL DATE LINE), this is a marginal sea belonging to the Southern Ocean. The sea, named for the British explorer SIR JAMES CLARK ROSS, who discovered it in 1841, indents the Antarctic coastline between Victoria Land (east of IDL) and Marie Byrd Land (west of IDL).

rostrum The upper jaw of a cetacean that forms the snout or beak (dolphin); or in fish, the forward projection of the snout; or in crustaceans (prawns), a forward-pointing spine of the carapace.

roundworm *See* NEMATODES.

ROV (Courtesy of NOAA/NURP)

ROV *(remotely operated vehicle)* In oceanography, an undersea vehicle equipped with one or more cameras designed for underwater exploration without a human operator aboard. The official development of the first ROV is vague, but credit should be given to Captain Luppis of the Austrian navy and the Scottish engineer Robert Whitehead, who between 1864 and 1868 developed the PUV (programmed underwater vehicle), the first real self-propelled torpedo powered by compressed air; and to the French engineer and pilot DIMITRI REBIKOFF, who, in 1953, developed the first tethered ROV named the Poodle, an improvement on his underwater scooter design the Torpille. The United States Navy deserves credit for advancing the ROV's capabilities in order to recover lost ordnance from the seafloor.

Royal Society of London Also labeled by some the British Royal Society or London Royal Society, this is the oldest scientific organization in the world. Headquartered in London, England, it was established in the year 1660 to promote the natural sciences and is still in existence today. The Royal Society played key roles in the founding and nurturing of the fields of oceanography, marine science, and marine biology.

rudder A flat, vertical plate or board used to steer a ship (or aircraft).

Sahul Banks The name given to a continental shelf system underlying the TIMOR SEA off the northern coast of Australia.

sail On a sailing vessel, it is the sheet of cloth or other material sewn from strips (for strength) and set on the mast in order to harness the wind, which, when managed properly, drives the vessel across the water. The mainsail is the largest and centermost sail. The jib is the forward triangular (lateen) sail, typically much smaller than the mainsail.

sailfish A tropical billfish and close relative of the marlin and swordfish. Sailfish are a target of sport fishing because of their fighting spirit and their unique beauty. Their saillike dorsal fin and upper body are dark blue; their bellies are a shimmering silver. Average length is six feet (1.8 m), though some may reach 10 feet (3 m). Sailfish belong to the family Istiophoridae, order Perciformes, class Actinopterygii, subphylum Vertebrata, phylum Chordata, kingdom Animalia.

Saint Lawrence, Gulf of A body of water belonging to the Atlantic Ocean centered at about 48° N and 62° W within the southeast coast of Canada.

Saint Peter and Saint Paul In the early days of navigation, this group of seamounts was an important landmark protruding above the surface in the mid-Atlantic Ocean. They lie in open waters between the eastern coast of South America and the westernmost bulge of Africa, centered just north of the equator at about 29° W. Of the cluster, the highest rock extends about 77 feet (23 m) above the surface.

salinity A term used when measuring the mass of salt per unit mass of SEAWATER. The standard measure is in grams per kilogram of water.

salmon A cold-water ANADROMOUS FISH, of which there are different members, Pacific and Atlantic. All salmon have varying shades of silver while at sea, where they feed on crustaceans and small fish, and become more brightly colored when spawning in freshwater. The largest salmon is the Pacific king, or Chinook, which averages between 40 and 50 pounds (18–23 kg) but has been known to grow to 110 pounds (50 kg). Salmon are an important food fish not only to larger marine life but to humans as well. Entire industries are based solely on salmon populations. In America, propagation is carefully managed and harvests are regulated; however, escaped Atlantic farmed salmon are thought to cause damage in Pacific stocks. Salmon belong in the family Salmonidae, order Clupeiformes, class Osteichthyes, subphylum Vertebrata, phylum Chordata, kingdom Animalia.

salps Free-swimming, barrel-shaped TUNICATES found living as floaters in tropical seas. They can be either solitary individuals or elongated colonials.

salt flats An area of land formed after the evaporation of salt water, below which it once resided. A salt flat may be the result of a bay that became enclosed by coral reef growth, or the buildup of a barrier island, or the subsidence of an already shallow sea. The Bonneville Salt Flats in Utah, United States, were once marine, and the DEAD SEA, which lies along the border between Israel and Jordan, is an example of a current building salt flat.

salt marsh A coastal area of brackish water dominated by communities of salt-tolerant plants (HALOPHYTES and GLYCOHALOPHYTES) that typically results from the buildup of silt as river water flows out to sea.

salts A chemical compound precipitated through the evaporation of seawater.

sand Rounded or angular mineralized particles with a diameter of 0.06–2.0 mm in size. Sand is smaller than gravel but larger than SILT or CLAY. *See also* MUD.

sand dollar A marine invertebrate related to the starfish, characterized by having a flat, disk-shaped shell (called a test) and many small perforations on one side that form a petallike or starlike design. Most sand dollars average about three inches (7.6 cm) in diameter and reside abundantly on sandy bottoms on both Atlantic and Pacific coasts. They move about on tube feet, which also serve to nourish. Sand dollars belong to the order Clypeasteroida, class Echinoidea, phylum Echinodermata, kingdom Animalia.

Sargasso Sea An area of open ocean in the North Atlantic occupying about 2,000,000 square miles (5,180,000 km^2) that lies roughly between the West Indies and the Azores and from about latitude 20° N to latitude 35° N, in the HORSE LATITUDES. This region is hydrographically bounded by the NORTH ATLANTIC, CANARY, NORTH EQUATORIAL, and ANTILLES CURRENTS and the GULF STREAM. It is distinguished from these surrounding waters by the rich abundance of marine plants that grow there, mainly because of the slowness of the ocean currents within. The Sargasso is steeped in early myth, played out by the imaginations of poets and writers who fancifully described sea monsters and the wrecks of ghost ships that could be seen from a distance, huddled in the center of a great strangle of seaweed. Today it is considered prime territory for marine biologists.

sargassum The name given by early sailors to all the various species of seaweed found floating atop the waters of an open area of the North Atlantic ocean. Thereafter, that region was known as the SARGASSO SEA. The original Portuguese word, *sargaço*, means grape; the name was used because the air bladders of some of the marine plants resemble grapes.

scales On fish, the protective, overlapping bony plates that cover part or all of the body. Some fish, such as catfish, hagfish, and lampreys, are scaleless, or "naked."

See also PLACOID; CTENOID; CYCLOID; COSMOID; and GANOID SCALES.

scallop A marine bivalve mollusk related to the clam and oyster. Scallops range worldwide from shallow littoral zones to ocean depths as great as 13,000 feet (3,962 m). Also called a fan shell or pecten, their shell has radiating ribs and winglike projections on the sides at the hinge. Scallops are filter feeders that range in size from one to six inches (2.5–15 cm) in length. Depending on type, they either are sessile or move quickly by snapping their shells; the outer edge of their mantle has tactile projections and many little blue primitive eyes that can be seen only if the animal remains undisturbed as it lies with its shell partly open. About 400 species are known. Scallops belong to the families Pectinidae, Entoliidae, and Propeamussiidae; order Ostreoida; class Bivalvia; phylum Mollusca; kingdom Animalia.

scarp In oceanography, it is an underwater slope that falls away sharply, usually separating plateaus or regions of moderate slope.

scavenger An animal that lives off decaying matter or other animals it finds already dead.

Schizophyta The bacteria. Schizophyta is a phylum under the kingdom Monera.

schooling Congregate behavior found in fish, in which a group of the same species live together, moving, mating, feeding, and defending the population from the depleting effects of predators.

schooner A sleek, fast sailing vessel with multiple masts, anywhere from two to seven, rigged fore and aft with lateen sails. These ships have been put to work in various ways, including being used for many voyages of exploration, and were popular in the late 1800s in the United States and Canada for their speed, comparatively small crew requirement, and ability to "lie closer to the wind" than a ship rigged with square sails. Today's schooners are pleasure boats, typically equipped with a sailing auxiliary (engine).

scope As a nautical term, it is used to describe the ratio of the length of an anchor line, from the vessel's bow to the anchor, to the water's depth.

scorpion fish Also called rockfish, this is a diverse group of fish with perchlike bodies and sharp, spiny, often highly poisonous projections that support the dorsal fin. Lacking a proper swim bladder, their hundreds of species range worldwide, inhabiting shallow, rocky waters but are found most abundantly in the Pacific Ocean. They have camouflage coloring, large heads, and wide gill openings and average about 12 inches (30.5 cm) in length. These fish feed on mussels, other fish, and crustaceans. Scorpion fish belong in the order Scorpaeniformes, class Actinopterygii, subphylum Vertebrata, phylum Chordata, kingdom Animalia.

Scotia Ridge An undersea, hairpin-shaped range linking the southern tip of South America with the northern tip of the Antarctic Peninsula. Its bend, bordering the undersea South Sandwich Trench, is centered at about 60° S and 25° W.

Scotia Sea A sea of the South Atlantic and Southern Oceans centered at about 58° S and 50° W, northeast of the Antarctic Peninsula, southeast of the southern tip of South America. Its area is defined within the Falkland, South Shetland, South Georgia, and South Sandwich Islands and lies within the path of the Drake Passage, which connects the waters of the Atlantic with those of the Pacific.

screw (Courtesy of NOAA)

screw Term used to describe a vessel's propeller. Marine propellers range anywhere between 10 inches and more than 25 feet (25 cm–7.6 m). Navy ships often have a three-blade screw and merchant ships, a four-blade screw.

Scripps Institution of Oceanography *(Scripps)* An oceanographic institution founded in 1903 as an independent biological research organization. Today it is devoted to graduate instruction and research in the marine sciences. Located on the coast of La Jolla, California, it became part of the University of California in 1912 and in 1925 was given its present name.
See also Useful websites in the Appendixes.

scuba *(self-contained underwater breathing apparatus)* *See* AQUA-LUNG.

sculpin A both marine and freshwater fish with a large, armored head; large mouth; and short, tapering body riddled with prickly spines around its head and fins. These otherwise scaleless fish are found worldwide near rocky shores, existing on small sea animals and bait from your hook. By most humans, sculpins are not considered a food fish because of their profusion of bones. Sculpins make up the family Cottidae, order Perciformes, class Osteichthyes, subphylum Vertebrata, phylum Chordata, kingdom Animalia.

scutes Large thick scales that serve as a protective armor, such as those covering all sea turtles, save the leatherback.

sea Speaking in specifics, a sea is a smaller, and typically shallower, subdivision of an ocean. Often they are marginal, meaning they are bordered by land and ocean, such as the CARIBBEAN SEA; or they may be partially enclosed, as is the BLACK SEA.

sea anchor A drag device, typically canvaslike, thrown overboard into the sea used to retard a boat's drift before the wind.
See also DROGUE.

sea anemone A marine polyp related to the hydras, corals, and jellyfish with a cylindrical body and tentacles adorned with nematocysts (stinging cells) that, along with its vivid color, give it the appearance of a flower. Anemones are typically fixed to rocks and such but are capable of slow movement. They subsist on small marine animals by paralyzing them and then pushing the victim into their central mouth. Sea anemones belong in the subclass Zoantharia, class Anthozoa, phylum Cnidaria, kingdom Animalia.

seabird A bird dependent on the sea for food, such as the auks, petrels, cormorants, and the albatross.

sea canyon An undersea canyon that intersects the rim of a continental shelf. Sea canyons are an important factor contributing to healthy bioactivity because of the regular, nutrient-rich upwellings that channel surfaceward from between their walls.

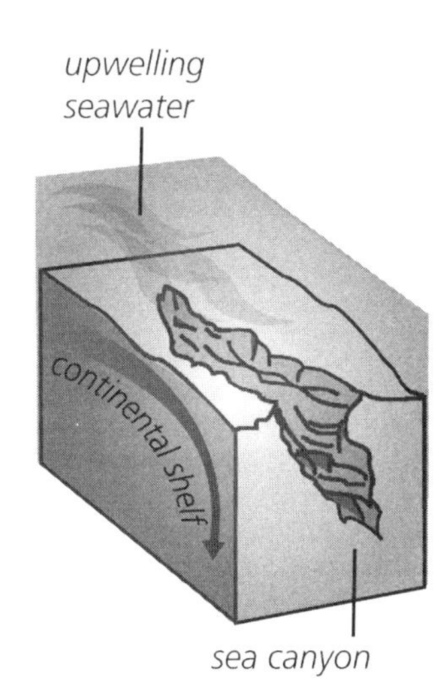

sea canyon

sea cow *See* SIRENIAN.

sea cucumber A marine invertebrate that gets its name from its torpedolike resemblance to a cucumber. Sea cucumbers are echinoderms, as are the starfish and sea urchin; however, their spines are less evident. They live in all oceans at all depths, and although most average about one foot (30.5 cm), some warm water varieties can grow as large as two to three feet (61–91 cm) long. The mouth is located at one end surrounded by tentacles that can be lengthened to catch food. When attacked or irritated, many can eject most of their body parts, only to grow new ones later. Sea cucumbers belong to the class Holothuroidea, phylum Echinodermata, kingdom Animalia.

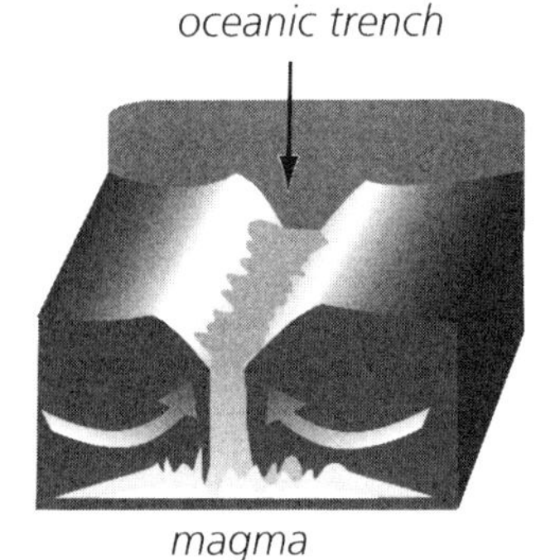

seafloor spreading

seafloor spreading Theory that the upwelling of magma from undersea volcanic ridges causes the ocean basins to widen, forming new seafloor, and therefore the continents to drift apart.
See also CONTINENTAL DRIFT; MAGNETIC ANOMALY.

sea grass A general term used for flowering plants, such as EELGRASS, growing in shallow marine waters and soft sediment.

sea horse A small marine fish that earned its name through the horselike appearance of its head and neck. They live in shallow, usually tropical, waters but are not good swimmers and attach themselves to seaweeds with their prehensile (wrap-grasping) tail. Most average less than six inches (15 cm) in length but the Pacific sea horse can grow to one foot (30.5 cm). Sea horses are remarkable in that, amid their combined utterance of musical sounds, the female deposits eggs in the male's pouch; afterward, the female departs and the male fertilizes them and carries them around until the young sea horses are ready to emerge. Sea horses belong to the family Syngnathidae, order Gasterosteiformes, class Osteichthyes, subphylum Vertebrata, phylum Chordata, kingdom Animalia.

sea horse (Courtesy of NOAA)

sea ice Term used for any ice formed from ocean water, whether it is free-floating or held fast to the shoreline. The latter is called fast ice.

seal A carnivorous marine mammal with a streamlined body, making it a fast swimmer. Seals have four flippers, or fin-feet, and big shiny black eyes, and some make doglike barking sounds. True seals, the earless seals (Phocidae), lack an external ear flap, though they have inner ears. Eared seals are known as sea lions (Otariidae). Most seals live in the Northern Hemisphere, but they can also be found in parts of the Southern Hemisphere. They feed on fish and shellfish, and most species are capable of using echolocation (sonar). Seals belong in the suborder Pinnipedia, order Carnivora, class Mammalia, subphylum Vertebrata, phylum Chordata, kingdom Animalia.

Sealab A saturation diving project sponsored by the U.S. Navy that first sprang to life in 1965. It was designed to prove that humans could exist at great depths under the sea. At one point during its use, the underwater Sealab habitat was occupied for more than one month, as the residents were supplied with a breathable mix of oxygen and helium.

sea lettuce A species of bright green marine alga found worldwide on rocks and shorelines. Ulva, as it is called, is high in vitamins A and B and can grow so rapidly it sometimes causes a bloom called a green tide.

sea level Term used to describe the average (mean) level of the sea surface, taken from the midpoint between the mean high and mean low tide marks.

sea lion A PINNIPED characterized mainly by its external ear flaps, as opposed to a SEAL, which possesses internal ears and is considered the true seal.

seamount A pointy undersea, usually volcanic mountain that rises to at least 3,300 feet (1,006 m) above the seafloor. Smaller submarine volcanoes are called sea knolls.

sea otter A playful aquatic member of the weasel family that can be found around kelp beds in the North Pacific Ocean. Most otters are freshwater animals; however, there is one marine species, Enhydra lutris. The sea otter likes to swim on its back and in this manner it carries its young on its belly and eats by cracking open shellfish, commonly by using a rock. Once hunted to near-extinction, they are now a protected species. Sea otters belong to the species *lutris*, genus *Enhydra*, family Mustelidae, order Carnivora, class Mammalia, subphylum Vertebrata, phylum Chordata, kingdom Animalia.

sea pen Relative of the jellyfish, it is a marine organism of the soft coral group that resembles a feather. There are about 300 species found almost worldwide. Sea pens form colonies where on a central stalk, called a peduncle (primary polyp), which is "rooted" in the mud or sand, many branches grow (secondary polyps). Many types are bioluminescent. Some can reach more than two feet (61 cm) in length. Sea pens belong in the order Pennatulacea, class Anthozoa, phylum Cnidaria, kingdom Animalia.

sea slug *See* NUDIBRANCH.

sea squirt A marine organism that resembles a lumpy potato. *See* TUNICATE.

sea star *See* STARFISH.

sea state A term used to describe the condition of the surface of the sea due to the effect of the wind and the waves.

sea turtle Any of several species of turtle adapted for swimming by having four large flippers as legs, no toes, and a lightweight shell. These animals range in size according to species and live almost constantly in the water, but the females go ashore to lay their eggs. The leatherback, an omnivore, is the largest sea turtle, averaging about 6.5 feet long (two meters) and 1,100 pounds (500 kg). The hawksbill, also an omnivore, is the smallest and averages 30 inches (76 cm) in length and around 100 pounds (45 kg) in weight. With the exception of the leatherback (family Dermochelyidae), sea turtles belong in the family Chelonidae, order Chelonia, class Reptilia, subphylum Vertebrata, phylum Chordata, kingdom Animalia.

sea urchin A marine organism characterized by the movable spines all over its roundish little body. The spines grow from its outer shell, or test, and on its flattish underside around its mouth. It has five sets of tube feet, which, in typical echinoderm fashion, it uses for locomotion (in conjunction with its spines). Fossils date back to 500 million years ago, in the Ordovician period. Sea urchins are found in both tropical and temperate waters and can be dark red, blue, or purple. They belong in the subclass Regularia, class Echinoidea, phylum Echinodermata, kingdom Animalia.

See also ARISTOTLE'S LANTERN.

seawater Seawater, or ocean water, is known as salt water. Depending on the area of the globe, the exact composition of seawater can vary, but on average it is 54 percent chlorine, 31 percent sodium, 1 percent potassium, 1 percent calcium, 3 percent sulfur, 4 percent magnesium, and 6 percent trace elements. The salinity of seawater is measured in terms of total dissolved salts in parts per thousand parts of water.

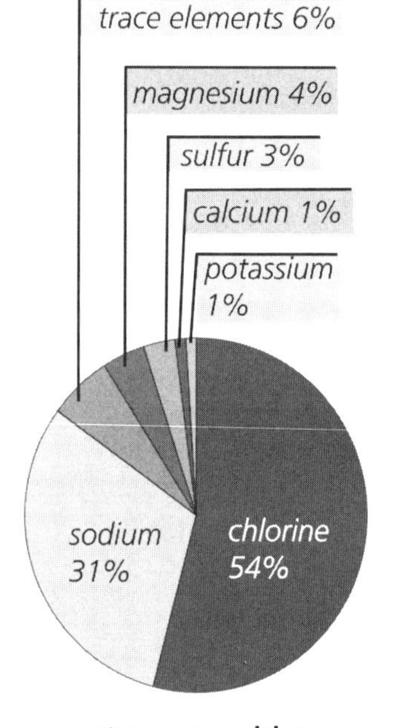

seawater composition

seaweed A word generalizing plants that live in the ocean.

Seaworld A division of the SAAMBR (South African Association for Marine Biological Research), a nongovernment, nonprofit marine conservation organization that also incorporates the ORI (Oceanographic Research Institute) and Sea World Education Center.

See also Useful websites in the Appendixes.

secretion Substances such as saliva, mucus, tears, bile, or hormone produced from cells or glands. Not a form of waste.

sediment Fine or coarse particles of matter that have settled to accumulate on the seafloor. The study of ocean sediment was one of the primary functions of the famous Challenger expedition of 1872–76.

seiche In lagoons, bays, or other types of enclosed or semienclosed bodies of water (and some areas of open ocean), it is the occasional rise and fall of water, unrelated to tidal action, caused by strong winds and/or changes in barometric pressure.

seine A large type of commercial fishing net deployed from a ship that has an appropriate mesh size enabling only the targeted fish species or larger species to be caught in the net.

seismology The scientific study of earthquakes.

semidiurnal tide A tide that has a period or cycle every half-day, meaning two highs and two lows in a 24-hour period. This is the predominant type of TIDE throughout the world, as opposed to a DIURNAL TIDE.

sessile Applies to sedentary animals permanently fixed to a substrate, such as a clam or mussel; not free-moving.

set A nautical term used to describe the direction toward which the current is flowing.

sextant An aid to navigation containing a graduated 60° arc, used by seagoers while away from land to measure the altitudes of stars, including the Sun, to determine longitude and latitude.

sexual dimorphism The apparent differences in males and females of the same species, such as their differing sizes and colors in relationship to one another.

sexual reproduction The generation of nonidentical organisms involving the fusion of gametes by two parents of opposite gender.

shallow-water waves Waves whose wavelength is at least 20 times the depth of the water beneath.

shark A marine CARTILAGINOUS FISH, of which there are about 250 species. Most sharks are predatory carnivores, but some, such as the largest member, the whale shark, at about 50 feet (15 m) in length, and the nearly as large basking shark, at about 40 feet (12 m) in length, are gentle plankton eaters. The smallest is the pygmy shark, checking in at about two feet (61 cm) in length. Shark species are found typically in warm waters, but some dog sharks exist in cooler temperate regions. Sharks form the order Selachii, class Chondrichthyes, phylum Chordata, kingdom Animalia.

See also GREAT WHITE SHARK; PLACOID SCALES.

shelf break The edge of an underwater cliff, the edge of a continental shelf, or the top of a coral slope where the angle changes sharply with descent.

shelf sea *See* EPICONTINENTAL SEA.

ship Term used to describe a vessel of considerable size, as opposed to a boat, which is regarded as a much smaller vessel. The ship is one of the oldest and most significant means of transportation in the world.

shipworm (Photograph by Scott McCutcheon)

shipworm A marine MOLLUSK also known as the teredo, which is not a worm at all, but an elongated clam. These are burrowing bivalves, the longest extending to around two feet (61 cm) into its coated burrow, though its shell remains about 0.5 inch (1.3 cm) long. Shipworms are known for the extensive damage they inflict on wooden structures such as docks, pilings, and the hulls of wooden ships. Shipworms belong to the family Teredinidae, order Eulamellibranchia, class Bivalvia, phylum Mollusca, kingdom Animalia.

shoal A submerged area of built-up sedimentary material along shallow coastal regions that, if not properly charted, can cause navigational problems for ships.

shore Term used to describe the general area of the coast between the low and high tide lines.

shoreline *See* TIDE LINE.

shrimp A marine shellfish related to the crab and lobster, sometimes referred to as a prawn. This crustacean has 10 jointed legs and a thin exoskeleton that sheds as it grows; unlike the crabs and lobsters, it is a swimmer. Shrimp have many colors, such as red, pink, gray, brown, white, and striped, and range in size from less than one inch (2.5 cm) in marine types to more than 12 inches (30.5 cm) in freshwater types. In the consumer industry, they are the most desired of the crustaceans. Shrimp belong in the order Decapoda, class Malacostraca, subphylum Crustacea, phylum Arthropoda, kingdom Animalia.

shroud On a sailing vessel, it is a thick rope or cable support "permanently" secured from the mast to the vessel's side.

See also STAY.

Sibuyan Sea A sea contained within the Philippine Islands, centered at about 13° N and 123° E surrounding the island of Sibuyan.

Sicily, Strait of Centered at about 37° N and 12° E, this is a region of the MEDITERRANEAN SEA separating Sicily, to the east, from Tunisia, Africa, to the west.

Sierra Leone Basin A basin of the eastern mid-Atlantic Ocean centered at about 5° N and 15° W off Sierra Leone, in western Africa. To the south is a curvature of the undersea Mid-Atlantic Ridge and to the west the undersea Sierra Leone Rise. The point of greatest depth averages roughly 16,500 feet (5,029 m).

siliclastic Type of SEDIMENT formed of silicate materials and rock fragments, such as mudstone and sandstone.

silicoflagellates Single-celled photosynthetic PHYTOPLANKTON whose glassy shells (frustules) are made up of tubular, rod-shaped elements. These algae first appeared in the Lower Cretaceous period and still exist today, though in far smaller numbers. Their skeletal remains make up 1 to 3 percent of MARINE SEDIMENT. These organisms belong to the order Dictyochales, phylum Chrysophyta, kingdom Plantae.

silt SEDIMENT formed of particles ranging between 0.01 and 0.05 mm in diameter, which is larger than CLAY, yet smaller than sand. MUD also lies within this range.
See also SAND.

siphon A tube through which bivalves breathe, filter-feed, and excrete.

Sipuncula Phylum of the peanut worms, which are cylindrical and have a frilly, tentacled mouth. These fall within the kingdom Animalia.

sirenian The sea cow, which is the ancient relative of the living dugong (whose habitat is more strictly marine) and the manatee (found in coastal rivers and lagoons). This marine mammal is a large, gentle, shy herbivore found in tropical waters. They are called sea cows because they graze on marine grasses and other water plants. Their gray fusiform bodies are heavy, with weak forelegs for flippers, no hind legs, and a cetaceanlike tail. They are hairless except for whiskers around the cleft upper lip. Because of their size, eight to 14 feet long (2.44–4.3 m), and 440 to 1,300 pounds (200–590 kg), depending on species, sea cows have few natural enemies. The living sirenians are classified in the families Trichechidae and Dugongidae, order Sirenia, class Mammalia, subphylum Vertebrata, phylum Chordata, kingdom Animalia.

Skagerrak A body of water centered at about 58° N and 9° E that separates Norway, to the north, from Denmark, to the south. By way of the KATTEGAT and subsequently of subordinate straits and channels, it connects the waters of the BALTIC SEA with those of the NORTH SEA.

skate A marine cartilaginous fish in the order of the RAYS.

slack water A brief interval between the shifting of tides when there is very little movement of the water. When navigating a small craft through certain regions within a network of ocean channels, especially in bad weather, it is sometimes best to wait for times of slack water to make the attempt.

sloop A vessel with a single mast and sails, main and jib, set fore and aft.

snail A GASTROPOD MOLLUSK with a single shell and single muscular foot on which it creeps slowly about. Aquatic snails possess gills. Some land-dwelling species have returned to the sea and consequently need to surface for air.

snorkel A hollow, J-shaped tube, open at the top with a mouthpiece attached to a swim mask at the other end, used for underwater skin-diving employing a breath-holding method.

SOFAR *(sound fixing and ranging)* Cylindrical-shaped unit that floats freely at a predetermined depth beneath the waves. These devices send acoustic signals to listening stations, allowing scientists to measure ocean circulation.

soft coral CORAL related to the gorgonians whose calcareous skeleton is embedded in the body wall. Typically they resemble feathers. Common species are the sea pansy, sea pen, organpipe, and whip coral. Soft corals predominate in the Indian and Indo-Pacific Oceans.

solar tide A daily partial tide caused solely by the gravitational force of the Sun. In the second century B.C.E., the Chinese recognized the relationship of the Moon and the Sun to the Earth's tides.

Solomon Sea A sea of the midwestern Pacific Ocean centered at about 7° S and 153° E. The Bismarck Archipelago borders it to the north, the CORAL SEA to the south, Papua New Guinea to the west, and the Solomon Islands and Pacific waters to the east.

solstice The point in Earth's orbit where the Sun appears from Earth to reach its highest and lowest declinations. The highest declination north occurs around June 21, marking the beginning of summer in the Northern Hemisphere, of winter in the Southern. The highest declination south occurs around December 22, marking the beginning of winter in the Northern Hemisphere, summer in the Southern.

solstitial tides Tides occurring near the times of the solstices. At these times, tropical tides may become unusually high.

solubility The extent to which a substance dissolves in a liquid, usually water.

solute A substance dissolved in a solution. In respect to seawater, salt is the solute.

solution A combination of two substances in which one is usually a solid (solute) and the other a liquid (solvent).

solvent A liquid containing a dissolved solute(s).

Somali Basin A basin of the northwest Indian Ocean centered along the equator at about 55° E. It lies off the east coast of Africa, near Somalia, and the point of greatest depth is roughly 16,750 feet (5,105 m) near the Owen Fracture Zone.

sonar *(sound, navigation, and ranging)* This system is similar to RADAR, except that it uses ECHOLOCATION (orientation by sound waves and their echoes), instead of radio waves. In nature, animals such as dolphins and bats use this form of this sensory system. In science, sonar is used for navigation in shipping and in fields such as acoustic tomography and acoustic oceanography.

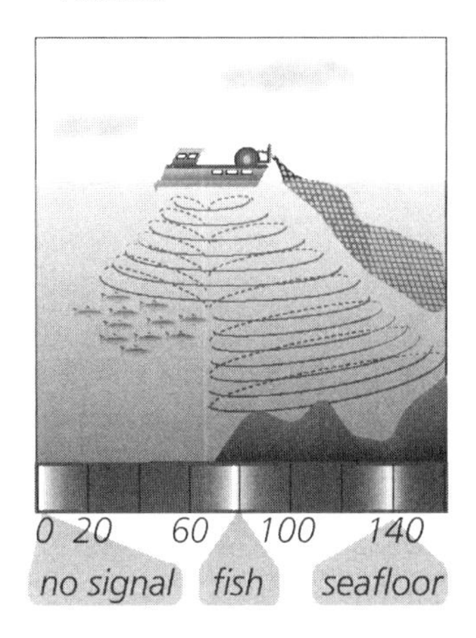

sonar

sounding An electronic means of ECHOLOCATION used to map or otherwise locate the seafloor by using sound waves.

South Atlantic Current A current of the western South Atlantic Ocean setting eastward toward Cape of Good Hope in South Africa, where, when combined with the Indian Ocean's southwesterly traveling AGULHAS CURRENT, it gives rise to the northwest flow of the BENGUELA CURRENT.

South Australian Basin A major basin of the East Indian Ocean centered at about 42° S and 130° E, south of Australia. Its point of greatest depth is roughly 19,200 feet (5,852 m) off the coast near Kangaroo Island, Australia.

South China Sea A marginal sea centered at about 12° N and 115° E that extends along the southeast coast of Asia between the Indian and Pacific Oceans. The Indonesian islands border it to the south, China to the north, the Philippine Islands to the east, and Vietnam and the Gulf of Thailand to the west.

Southeast Pacific Basin A large basin of the Pacific Ocean centered at about 60° S and 100° W between the southwestern tip of South America and the undersea East Pacific Rise. The deepest point is roughly 16,900 feet (5,151 m) in the Humboldt Plain.

South Equatorial Current An equatorial current flowing westward and south of the equator in the Atlantic, Pacific, and Indian Oceans. It occurs immediately south of the Equatorial Counter Current (which flows eastward) in the Atlantic.

Southern Hemisphere The half of the Earth south of the equator.

Southern Ocean Also sometimes known as the Antarctic Ocean, it is one of the five divisions of the WORLD OCEAN. Its surface area covers

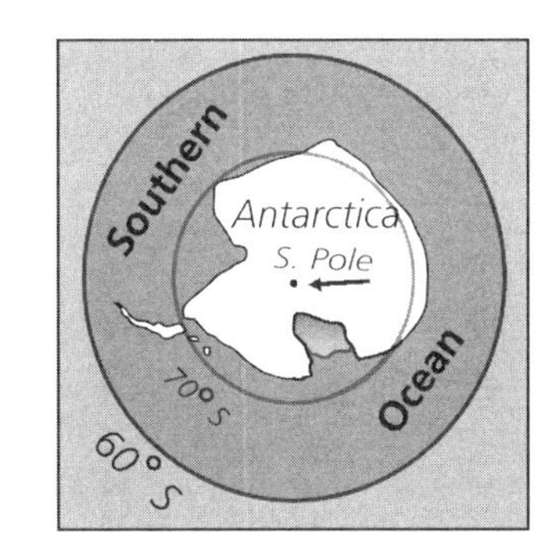

Southern Ocean

7,848,000 square miles (20,327,000 km^2) from the coast of Antarctica northward to 60° S, where its boundaries merge with the Pacific, Indian, and Atlantic Oceans. The Southern Ocean, though cold, supports sizable plankton and krill populations that make up part of the food chain for seabirds, fish, invertebrates, and cetaceans.
See also ARCTIC OCEAN; ATLANTIC OCEAN; INDIAN OCEAN; and PACIFIC OCEAN.

South Indian Basin Centered at around 60° S and 110° E, this is one of four oceanic basins encircling the continent of Antarctica; the others are the BELLINGSHAUSEN, ENDERBY, and WEDDELL ABYSSAL PLAINS. Depths average 14,800 feet (4,511 m).

South Pole Geographically, the South Pole sits at latitude 90° S on the Antarctic continent. It is the southern point of the Earth's axis and is distinguished from magnetic south. In 1911 the Norwegian explorer ROALD AMUNDSEN was the first European to reach the South Pole.

Southwest Pacific Basin A large basin of the Pacific Ocean centered at about 40° S and 150° W. The area is defined by the undersea features of the NORTHEAST PACIFIC BASIN to the north, the Pacific-Antarctic Ridge to the south, the East Pacific Rise to the east, and the Tonga and KERMADEC TRENCHES as well as New Zealand to the west. Deepest points average about 5,393 feet (1,644 m).

spanker On a sailing vessel, the fore-and-aft sail on the mizzenmast.

spar A nautical term sometimes used generally to describe masts, yards, booms, and other structures. However, more specifically a spar is the pole attached to the mast that supports the sail.

species In TAXONOMY, a classification group consisting of closely related genera, below the level of a genus.
See also KINGDOM.

species diversity The number of different species present and their relative abundance in a community.

spicule A spiky, minute calcareous or siliceous skeletal support in sponges, radiolarians, soft corals, and sea cucumbers.

spiricals An opening or pair of openings found in many benthic organisms, such as arthropods and cartilaginous fishes, that acts as a ventilation system, allowing water to be drawn in and channeled to the gills.

spit A low-lying sandbank with one end attached to land and the other reaching out to sea. Spits typically result from the action of longshore currents.

splash zone Coastal area above the high tide line affected by saltwater spray.
See also SUPRALITTORAL ZONE.

sponge An aquatic organism of which there are thousands of brilliantly colored species found worldwide at a wide range of depths. Most species are marine, but a few are freshwater. Sponges are most abundant in tropical waters, ranging in size from about 0.5 inch to five feet (1.3–152 cm) and characterized as having a porous structure and a siliceous, calcareous, or proteinaceous skeleton of thorny, interlocking fibers called spongin, which is related to horn. They are sessile filter feeders and appear so sedentary that before the 18th century they were categorized as plants. Since ancient times sponge skeletons, because of the fineness of their mesh, have been used to hold water for bathing or drinking. Sponges make up the phylum Porifera in the kingdom Animalia.

spore Minute, unicellular asexual reproductive structure of an alga.

Sporozoa and Microspora Phyla of the parasitic protozoans, within the kingdom Protozoa.

spreading ridge Undersea mountain ridge built up by magma that flows in to fill gaps between widening lithospheric plates, a phenomenon that causes the seafloor to spread as new material forms.
See also SEAFLOOR SPREADING.

spring tide A maximal TIDE, either the highest of the high or the lowest of the low, occurring semimonthly as the result of the Moon's being new or full.

spyhopping Behavioral habit of cetaceans, which raise the head and eyes above the water to observe their surroundings. Compare to BREACHING.

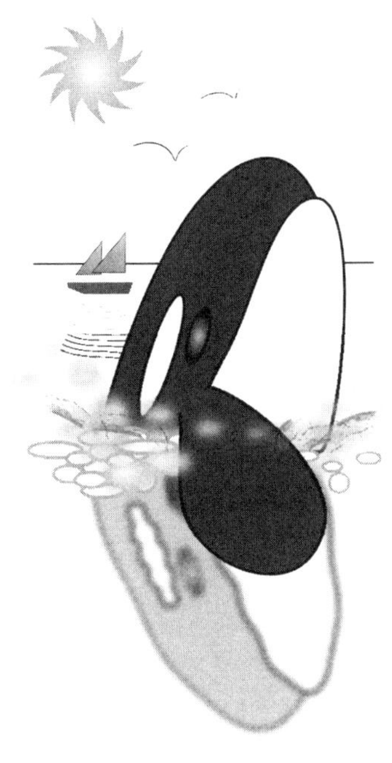

spyhopping

square rigger A sailing vessel bearing square sails, as opposed to lateen (triangular) sails.
See also LATEEN SAIL.

squid A 10-armed marine CEPHALOPOD mollusk, highly developed and a skillful predator.
See also CHROMATOPHORES; CUTTLEFISH; GIANT SQUID.

stack Used to describe an offshore pillar of rock, eroded from its onetime attachment to land by the constant action of the waves.

standing stock Biomass (total weight) of a population present in any given area or volume at any given time.

starboard On a vessel, when facing forward it is the right side of the ship. Starboard is indicated by a green light aimed toward the bow, visible to all traffic but that which travels astern. Compare to PORT.

starfish (Photograph by Scott McCutcheon)

starfish A carnivorous marine echinoderm more appropriately known as the sea star, for it is not a fish. Sea stars can be found in rocky littoral zones of all seas and are characterized as having a flat body of radial symmetry with five arms (more in some species); they are usually dull yellow or orange but are also found in bright colors. They range in size from less than 0.5 inch to more than three feet (1.3–91 cm) in diameter. True to echinoderm nature, they move about on tube feet located on their undersides and subsist mostly on bivalve mollusks. The brittle star, a completely different class of sea star, has long, slender, jointed arms and is found in deeper waters. Sea stars are classified in the class Asteroidae, phylum Echinodermata, kingdom Animalia.

statocyst Sensory organ of equilibrium containing mineral grains (statoliths) that are stimulated in response to an animal's movements and help the animal sense the direction of gravity.

stay On a sailing vessel, the heavy working lines used to support the mast, suspended either from mast to mast or from mast to vessel. From the mast, fore and aft stays are those that lead forward; backstays are those that extend to the vessel's sides.

See also SHROUD.

stenohaline A term associated with aquatic organisms with limited tolerance of salt water.

stern The after, or rear, part of the boat; opposite of bow.

storm beach A beach with debris washed high ashore as a result of storm waves.

storm surge A phenomenon whereby the sea level in a particular region rises above normal as a result of heavy weather somewhere nearby over the ocean, such as a hurricane. Storm surges are caused by three factors: low barometric pressure, allowing the surface of the water to rise (called the inverse barometer effect); wind that pushes water to the shore; and correspondence with tidal motion.

strait A narrow strip of water that links two larger bodies of water.

strandline On a beach, it is the line where debris piles up on the shore from the normal action of waves.

stratification 1) For fluids, it is a layering according to density. Stable stratification occurs with the continuous decrease in density (uniform or not) with distance from the Earth's center. 2) In general, it is the deposition of substances into horizontal layers that may vary in composition or properties.

stridulation The producing of sound through a plucking action of a specialized part of the body. Lobster stridulation noises, such as growls, rasps, rattles, and pops, are derived from rubbing pads at the base of their antennae against ridges on their heads.

sturgeon A primitive aquatic fish; some species are marine, some anadromous, and some exclusive to freshwater lakes. These fish may be found in the northern regions of Europe, Asia, and North America. They have a long, narrow body; scutes (bony plates) along their back and sides; and feelers (barbels) hanging below the head that serve to locate food, such as crayfish, snails, larvae, and small fish, which they suck into their toothless mouth. The largest is the Russian sturgeon, reaching up to 13 feet (four meters). The smallest is the American shovel-nosed sturgeon, about three feet (91 cm). Sturgeons belong to the family Acipenseridae, order Acipenseriformes, class Osteichthyes, subphylum Vertebrata, phylum Chordata, kingdom Animalia.

subclass In TAXONOMY, a classification group consisting of closely related genera, below a class and above an order.
See also KINGDOM.

subduction 1) In oceanography, surface water that slips below into intermediate depths, causing the formation of water masses in the permanent thermocline. This is produced by the EKMAN TRANSPORT. 2) In geology, a phenomenon occurring between separate lithospheric plates of the Earth in which one slides under another. Usually the more dense oceanic plates are the ones that dive beneath the less dense continental plates.
See also PLATE TECTONICS.

sublittoral zone *See* INFRALITTORAL ZONE.

submerged beach A coastal region that has dropped below sea level as a result of a general rise in the water level or a drop in the shoreline.
See also RAISED BEACH RELICT BEACH.

submersible A small, portable underwater vehicle capable of performing remote operations or of carrying a research crew to great depths in the ocean, far beyond those attainable by free divers.
See also BALLARD, ROBERT D.; BEEBE, CHARLES WILIAM; ROV.

suboceanic province Term sometimes used to describe the benthic pelagic region of the ocean.

subsidence Used to describe land that has dropped into the sea, by whatever means, perhaps through a rise in sea level, volcanism, or collapse due to the weakness of the continental material under the force of water pressure.
See also BLUE HOLES.

submersible (Courtesy of NOAA/NURP)

subspecies Organisms distinguishable from other populations within the same species group because of minor evolutionary differences.

substrate The seabed where plants and animals may live. Substrates may be hard or soft, depending on their type.

subsurface current An underwater current differing from the surface current in speed (usually slower) and direction.

Subtropical Convergence Also called Subtropical Front, the broken water boundary that exists between the waters of the Southern Ocean and the subtropical waters that lie just to their north, defined by a difference in temperature and salinity.

Suez, Gulf of Centered at about 29° N and 33° E, this is one of two tongues of water that fork from the northernmost point of the RED SEA to form an aquatic boundary around most of Sinai. The Suez renders the western tongue, and the other, the Gulf of Aqaba, constitutes the eastern.

Sulu Sea Centered at about 9° N and 120° E, this is a body of water contained between the Philippines to the north and Malaysia to the south. To the south lies the CELEBES SEA; to the west, the SOUTH CHINA SEA.

supralittoral zone Also called the splash zone, this is the area of shore that is permanently exposed to the air, except for occasional wetting by sea spray or high waves. Examples of life in this zone are lichen and other glycohalophytes. Lower in the supralittoral fringe are barnacles and cyanobacteria (blue-green algae).

surf zone The area of the NEARSHORE ZONE extending from the inner breakers shoreward to the swash zone.

surface scattering layer In the WATER COLUMN, a surface layer of marine organisms that scatters sound, extending from the surface to depths as great as 600 feet (183 m).
See also DEEP SCATTERING LAYER.

surge channel A deep, narrow channel eroded from a hard substrate that allows the passage of water to and from more shallow areas of the ocean.

suspension feeders Organisms that feed by straining particles from the water around them. Examples are sponges, corals, even baleen whales.

swash zone The area of the NEARSHORE ZONE where the beach face is alternately covered by wave swash and then exposed by backwash.

swell A long, nonbreaking WAVE or succession of waves on the surface of the water that has moved away from its source of generation without interruption.

swim bladder A gas-filled cavity that acts as a float for fish (called the pneumatocyst) and for plants (called fucoid gas bladders) that keeps the organism buoyant as it lives within its water column.

symbiont Also spelled *symbiote,* an individual organism that shares a symbiotic relationship.

symbiosis Living together of two different organisms in intimate association; the arrangement often benefits both.

sympatric Pertaining to species or subspecies that occur together and are distributed within the same, or more often overlapping, geographical regions; opposite of ALLOPATRIC.

systematics The study of taxonomic classifications, the phylogeny of organisms, and all the associated language.

tablemount *See* GUYOT.

tack Tacking is a method used when a sailing vessel attempts to use wind to move in a direction different from the wind.

tactile Pertaining to the sensory perception of touch.

tail stock In cetaceans and sirenians, it is the area that connects the tail with the rest of the body.
See also CAUDAL PEDUNCLE.

tartan A small, single-masted Mediterranean ship rigged with a large lateen (triangular) sail.

Tasman Basin An oceanic basin between the Southwest Pacific and Indian Oceans centered at about 45° S and 158° E. It lies between New Zealand and southeastern Australia; its point of greatest depth is roughly 16,270 feet (4,959 m).

Tasman Sea Centered at about 40° S and 160° E, this is a body of water connected to the South Pacific and South Indian Oceans. New Zealand borders it to the east, Australia to the west, the CORAL SEA to the north, and the Southern (Antarctic) Ocean to the south.

taxis In a plant or animal, it is movement toward or away from an external stimulus such as light, sound, current, or physical touch.

taxon Any group of genetically related organisms to which a taxonomic category can be assigned, such as those within a species, genus, order, or phylum.

taxonomy The science of classifying living or once-living genetically related organisms and isolating them into various stages of family groups.

teleost Within the realm of evolution, it is the most advanced of the bony fishes. This is not a taxonomic classification but a general grouping of all types of modern fish, specifically those identified as having tail lobes of the same size and fusiform body shapes.

telestomes A category encompassing the ray-finned fishes and the lobe-finned fishes, associating them as "true" fishes, meaning those that possess a terminal mouth and bony skeleton.

temperate zone Mild, intermediate latitude zone of the Earth that experiences nominal precipitation in all seasons. The North Temperate Zone lies between the ARCTIC CIRCLE and the TROPIC OF CANCER, and the South Temperate Zone between the ANTARCTIC CIRCLE and the TROPIC OF CAPRICORN.

tentacle A flexible growth on invertebrates used for feeding or grasping.

teredo *See* SHIPWORM.

terrapin Also called the diamond back turtle. This animal is native to the coastal salt marshes of the eastern United States, from Massachusetts to Texas.

See also SEA TURTLE.

test The hard outer covering secreted by certain organisms, such as the radiolarian, sand dollar, and sea urchin.

Tethys Sea An extinct sea of the Mesozoic era that separated GONDWANALAND from LAURASIA. The CASPIAN and ARAL SEAS are remnants of the Tethys.

tetrapod Any four-limbed vertebrate animal, such as a reptile, mammal, amphibian, or some birds, that has wrists, ankles, toes, and fingers. It is not a taxonomic classification.

Thailand, Gulf of An arm of the SOUTH CHINA SEA centered at about 10° N and 102° E. Thailand borders it to the north, Cambodia and Vietnam to the east, the Malay Peninsula to the west; it is open to circulation to the south.

therapsids Extinct mammallike reptiles that existed during the end of the Triassic period. Mammals were thought to have evolved from these animals.

thermal Used as a noun, it is a rising current of heated liquid or air.

thermal spring On the ocean floor, it is a spring that releases water that is warmer than that which surrounds it.

See HYDROTHERMAL VENTS; VENT COMMUNITIES.

thermal stratification In the ocean, it is the layering of water of different temperature gradients.

thermocline In the ocean, the invisible boundary between the cold and warm layers and the depth below which the temperatures may begin to decline sharply.

thermohaline Descriptive of global oceanic circulation driven by variation in water temperature and saline content.

thermohaline circulation Oceanic circulation of the polar regions driven by seawater density, salinity, and temperature; density is determined by salinity and temperature. In thermohaline condition, nutrient-rich deep water rises to mix with wind-driven surface currents.

thermometer An instrument used to measure the degrees of heat or cold of gases, liquids, or solids in reference to a predetermined scale, whether it is Kelvin, Celsius, or Fahrenheit.
See also 1593; 1641; 1714.

thermosphere One of two zones (the other is the PSYCHROSPHERE) into which the depths of the sea are sometimes separated according to temperature. The thermosphere consists of water occupying the upper regions of the sea with temperatures greater than 50° F (10° C).
See also OCEAN TROPOSPHERE.

tidal bore Also called bore, this is a tidal wave with a high, well-defined crest that propagates up a relatively shallow, sloping estuary or river as a single wave. The bore's leading edge sweeps upstream, causing an abrupt rise in water level; it breaks continuously and is often accompanied by frequent undulations. Tidal bores are not common and are associated with the presence of a full moon (high spring tides), as well as being influenced by conical-shaped river entrances or entrances whose bottom rises quickly into a shoal.

tidal range The range of height difference measured between high tide and the next low tide.

tide line In respect to oceans and seas, it is the edge of the water as it moves up and down across the shore with the action of the tides.

tide rip An elongated mass of swirling, agitated water caused when tidal currents meet. A rip happens when currents unrelated to tides meet or when fast currents encounter an irregular or suddenly shallow bottom. Good fishing can be had along a rip.
See also OVERFALLS.

tides The regular and timely rise and fall of any body of water, large or small, due to gravitational interactions of the Sun, the Moon, and the Earth. On each end of their extremes, they are categorized as low and high tides and this phenomenon must be taken into consideration in nearly every activity associated with the ocean, especially that along

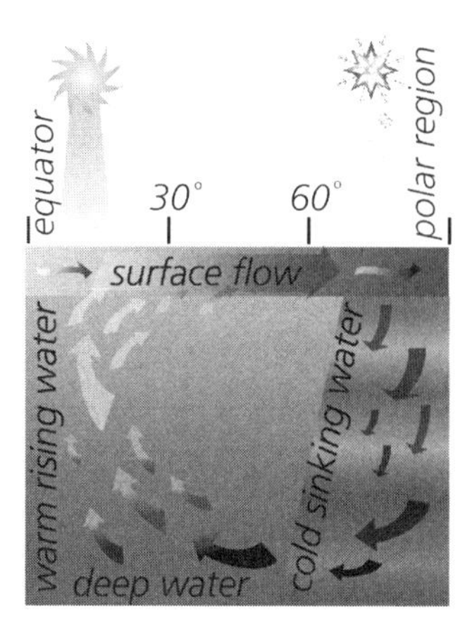

thermohaline circulation

tides (high) (Photograph by Scott McCutcheon)

tides (low) (Photograph by Scott McCutcheon)

the coast. Tides are classified as either diurnal (one daily height) or semidiurnal (two daily heights) and can be thought of in the simplest terms as "water that follows the Moon." Tides help to keep the coastlines clean of waste and debris, and their levels differ all across the world. Some areas exhibit very little tide swing, such as in the Mediterranean; other areas, such as Alaska, are subject to an enormous swing of sometimes up to a 30-foot difference between high and low.

tide tables Tables that provide daily information on an area's predicted high and low tide levels and their time of occurrence.

tiller On a boat, a bar or handle attached to and used to manipulate the rudder or outboard motor and thereby dictate the vessel's heading.

Timor Sea A marginal sea of the East Indian Ocean centered at about 11° S and 128° E. Australia borders it to the south, Indonesia to the north, the ARAFURA SEA to the east, and the INDIAN OCEAN proper to the west.

tombolo A small gravel or sandy bar connecting an island to the mainland or to another island.

topography On a map or chart, the detailed and precise configuration of geographic surface features.

totipalmate In some birds, webbing between the toes to facilitate swimming.

toxic Poisonous.

toxin Any poisonous substance of either plant or animal origin capable of producing a harmful effect on organisms, either by an immediate large dose or by small doses over a period. When toxin is present in certain dinoflagellates, RED TIDES result.

trace elements Chemical elements essential to life yet existing in amounts too tiny to list.

trade winds Also known as tropical easterlies, they are named from the early sailing days, when trade between peoples depended on the regularity of these winds to fill the sails and thus drive the ships to port. These winds diverge from subtropical belts of high pressure and center between 3° and 40° N and S toward the equator, north to east in the Northern Hemisphere and south to east in the Southern Hemisphere.

transform fault A break in the Earth's crust, characteristic of midoceanic ridges, where the line of the ridge is offset by the fault.
See also SPREADING RIDGE.

transom On a boat, the cross section of a square stern where an outboard motor, when present, is attached.

Transpolar Drift Arctic current that transports cold, freshwater north from the LAPTEV SEA, across the Eurasian Basin, into the Fram Strait, and thus to the Atlantic by way of the EAST GREENLAND CURRENT.

trawling A term used by both marine biologists and commercial fishermen. In fishing, it is the means by which the harvester gathers benthic product, such as shrimp or cod. Biologists use it in reference to the haul of SEDIMENT or other bottom material intended for study. Both means employ a net, of whatever mesh, that is pulled behind a vessel, at whatever depth, depending on the target population.

triangulation A trigonometric method used to pinpoint an object's position by taking bearings on two fixed points that are a known distance apart.

trilobite Looking like a flattened centipede, it is an extinct marine arthropod of the Paleozoic era, representing more than half of the known fossils from the Cambrian period. They disappeared during the Permian period. Paleontologists use their fossils in dating rock structures. The name *trilobite* refers to the furrows that run as a pair down the body, dividing it into three regions. These animals are most closely related to the chelicerates (arthropods with modified pincerlike appendages), which include the horseshoe crabs and spiders.

tropic of Cancer Latitude 23° north of the equator, marking the farthest limit (north) where the Sun can appear directly overhead. This name indicates the Sun's location within the constellation of Cancer when the event occurs.

tropic of Capricorn Latitude 23° south of the equator, marking the farthest limit (south) where the Sun can appear directly overhead. This name indicates the Sun's location within the constellation of Capricorn when the event occurs.

Tropics The regions that encompass 3,418 miles (5,500 km) of equatorial Earth between the TROPIC OF CANCER and the TROPIC OF CAPRICORN. Though there can be definite dry and rainy periods, temperatures do not vary greatly in the Tropics because of the lack of seasonal change. This is due to the amount of daylight, which remains basically the same year around.

trough The still-water depression between the crests of two waves.

true north *See* GEOGRAPHIC NORTH.

true wind The actual direction from which the wind is blowing, as opposed to an apparent wind, which is true wind coupled with wind derived from a vessel's own speed.

tsunami Pronounced "sue-naw-mee," is a rogue wave caused by an undersea volcanic eruption, earthquake, or landslide. Tsunami waves from 10 to 200 feet (three to 61 m) high have been recorded.

Tsushima Current A segment of the Kuroshio system, this is a North Pacific Ocean current setting northeast into the East China Sea and the Sea of Japan. The current branches as it encounters Japan's Tsushima Islands, with the western branch becoming stronger during summer and the eastern branch remaining weak year round.

tubenose Term used generally to describe seabirds that have tubular rather than open nostrils, such as the albatross and petrel.

tuna Also called tunnies, they are a well-developed pelagic marine game and food fish related to the mackerel. They have streamlined bodies that taper narrowly toward a deeply forked tail, two pectoral fins, and five additional finlets along the back. The smallest tunny is the false albacore, averaging about 10 pounds (4.5 kg). The largest is the great bluefin, averaging 200 to 500 pounds (91–227 kg) and sometimes up to 14 feet (4.3 m) and 1,500 pounds (680 kg); these granddaddies fetch a very high price in Japan. Tunas belong in the family Scombridae, order Perciformes, class Osteichthyes, subphylum Vertebrata, phylum Chordata, kingdom Animalia.

tunic In TUNICATES, it is the leathery or gelatinous covering secreted by the body wall.

tunicate A marine organism that, in cylindrical or globular adult form, is a sessile filter feeder found attached to rocks, pilings, or even boats. It is classified as a vertebrate, but that characteristic is evident only in the free swimming larva, which resembles a tadpole. The most familiar example of a tunicate is the sea squirt, whose soft body surrounded by a thick tunic averages about two inches (five centimeters) long. Other examples of tunicates are the salps and the larvaceans. Tunicates belong in the subphylum Urochordata, phylum Chordata, kingdom Animalia.

turbidity currents Ocean currents distinguished as being load-bearing currents present along continental slopes and often associated with undersea seismic activity or floods that result in rapid river flow into the sea. These currents can be responsible for underwater landslides and the reshaping of the sediment through the gravitational deposition of their contents.

turbulence The random and disorderly mixing of water in which velocity fluctuations distort and confuse the liquid flow.

Tyrrhenian Sea A body of water within the central MEDITERRANEAN SEA centered at about 40° N and 12° E. Italy borders it to the east, Corsica and Sardinia to the west, Sicily to the south, and the LIGURIAN SEA to the north. The Strait of Messina connects it with the IONIAN SEA.

ultraplankton Phytoplankton smaller than nanoplankton and difficult to separate from water.

under way As a nautical term, any vessel in motion. Even a disabled or abandoned vessel drifting with the current is deemed under way.

unicellular Single celled, such as the protozoan.

upwelling The slow rising of water to the surface from deeper areas of the ocean. These nutrient-rich waters are produced primarily by the action of surface winds. They carry important nitrates and phosphates and are usually colder than the surface water. The TRADE WINDS, along with the CORIOLIS EFFECT, can cause upwellings along the west coasts of continents. The colder water commonly lowers the air

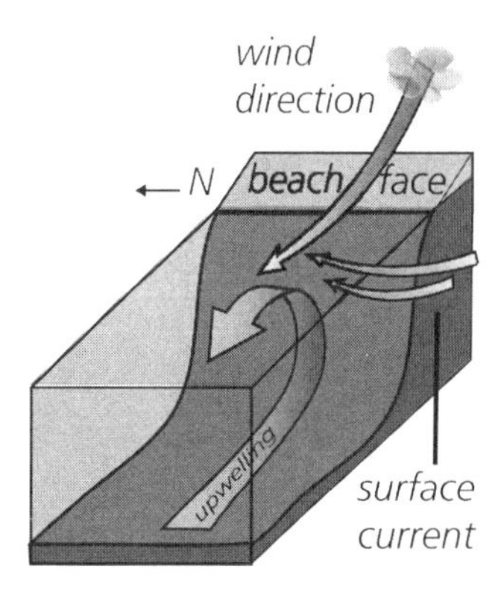

upwelling

temperature, increasing relative humidity and often producing fog.
See also DOWNWELLING.

vados water Pertaining to the Earth, it is the water existing above the permanent water table, or groundwater level.

vascular plants Plants possessing vessels that circulate and conduct fluid and therefore minerals around their parts.

veliger A larval MOLLUSK, in the stage when it develops its velum (swimming organ).

Vema Trench A deep channel in the North Indian Ocean important for allowing deep-water circulation between the Nares and HATTERAS ABYSSAL PLAINS. Average depth is roughly 21,000 feet (6,400 m).

Venezuelan and Colombian Basins Oceanic basins of the east CARIBBEAN SEA between South America and the Greater and Lesser Antilles. The deepest points average about 16,800 feet (5,120 m) in the Venezuelan and 14,880 feet (4,535 m) in the Colombian Basin.

vent communities Communities of living things near deep-sea HYDROTHERMAL VENTS whose members are completely nondependent on photosynthesis. Huge colonies of tubeworms thrive there within the pitch blackness, existing on bacteria. Large crabs, oysters, clams, fish, shrimp, and various other forms of biota also exist around these vents, which emit hot, mineral-rich waters. Early in the history of marine biology, life was thought to be nonexistent in these deep, unreachable areas.
See also BALLARD, ROBERT D.; THERMAL SPRING.

ventral fin In fish, it is the lower or belly fin.

vertebrate Animals ranging from fish to human who have a segmented spinal column and a distinct, well-differentiated head.

vertical migration Pertaining to marine animals, it is the daily or seasonal up and down movement between the photic zone and midwater depths of the water column.

vertical zonation Distinct separation of groups of organisms within different levels of the WATER COLUMN.

vessel A boat, ship, raft, or other means of water transport.

vibrissae Whiskers.

Visayan Sea Centered at about 11° N and 124° E, this is a small sea contained within the latticework of the Philippine Islands. The SIBUYAN SEA borders it to the northwest, the Samar Sea to the east, the BOHOL SEA to the south, and the SULU SEA to the west.

GLOSSARY

vados water – Visayan Sea

viviparous Female animal that gives birth to live young after it is nourished within her body.
See also OVOVIVIPAROUS.

volcanic island *See* ISLAND.

volcanism Volcanic activity.

vortex In water, a spiraling whirlpool that sucks anything within its limited range toward the center.

Wallace Line Imaginary biogeographical boundary line first introduced by the British naturalist ALFRED R. WALLACE, that marks a division between different yet related species that are separated by some kind of natural barrier, such as a large body of water.

walrus A marine mammal, the largest of the PINNIPEDS, distinguished by long ivory tusks and its fat cheeks bristling with whiskers. Walruses can dive but live mainly on ice floes and in shallow arctic waters, feeding on shellfish. They are fat, wrinkly, brown, and practically hairless, and males average about 10 feet (three meters) long and weigh up to 3,000 pounds (1,360 kg). Walruses belong in the family Odobenidae, suborder Pinnipedia, order Carnivora, class Mammalia, subphylum Vertebrata, phylum Chordata, kingdom Animalia.

Walvis Ridge An undersea mountain range in the South Atlantic Ocean centered at about 25° S and 5° E. It extends westward from the southwestern coast of Africa to connect with the undersea Mid-Atlantic Ridge. The highest peak is the Valdivia Seamount, which reaches within 3,040 feet (926 m) of the surface.

water column The vertical depth of water.

water cycle *See* HYDROLOGICAL CYCLE.

water flea A familiar freshwater crustacean that lives floating on lakes and slow-running streams and moves in quick bursts across the surface.

waterline Term used to describe the surface plane of the water that meets the atmosphere.

water mass A large body or selected volume of water that has a history of common origin and uniform characteristics of temperature and salinity.

water spout A column of water drawn up into a whirling spout above the surface of the water by a tornado style of wind that has formed above it.

water vapor Water heated to the point of becoming gaseous in the air, but not hot enough to form steam.

water column

air trapping system

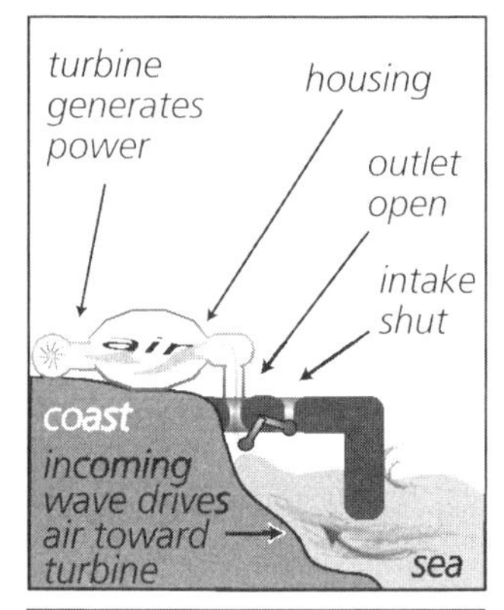

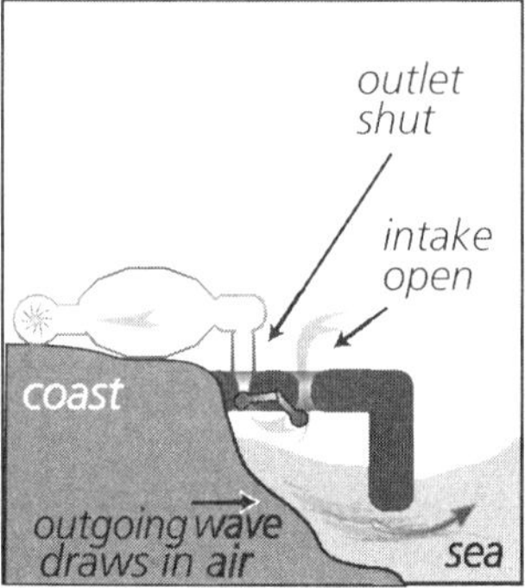

wave power

wave In the large bodies of water, it is a SWELL or curl of water, usually a regular procession of them, moving along the surface, away from the point of origin. Storm-generated waves may have crests miles apart. A groundswell is a long, storm-generated swell that serves as a warning that a storm is on its way.

wave height The vertical distance between a wave's crest and the preceding trough.

wavelength The horizontal distance between two similar points on successive waves, such as crest to crest or trough to trough.

wave period The elapsed time as two successive waves pass a stationary point.

wave power Electricity generated by using the energy of waves.

weathering The erosion of a coastline through the natural action of wind, waves, and water flow.

Weddell Abyssal Plain Centered at around 65° S and 10° W, this is one of four oceanic plains encircling the continent of Antarctica, the others are the BELLINGSHAUSEN and ENDERBY ABYSSAL PLAINS and the SOUTH INDIAN BASIN. Depths average around 16,800 feet (5,120 m).

Weddell Current A current of the Southern Ocean setting eastward from the WEDDELL SEA along the coast of Antarctica that is not a part of the WEST WIND DRIFT.

Weddell-Scotia Confluence A boundary of water forming from the Antarctic Peninsula northeastward into the southern SCOTIA SEA. The separation of the WEDDELL SEA and the Scotia Sea is determined by the temperature and salinity differences dictated by this confluence.

Weddell Sea A marginal, usually ice-covered sea of the Southern Ocean centered at about 72° S and 40° W between Queen Maud Land and the Antarctic Peninsula. It is named for the British navigator JAMES WEDDELL, who in 1823 claimed to be its discoverer.

weed drag A sampling device used to gather seaweeds from below the surface, such as the type that consists of a three-pronged piece of weighted metal attached to a length of cable and towed behind a vessel.

West Australian Current A branch of the Antarctic's WEST WIND DRIFT that sets northward when it encounters the western coast of Australia and continues as a coastal current into tropical regions.

westerlies Winds blowing from the west, found in both the Northern (blowing from the southwest), and the Southern (blowing from the northwest)

Hemispheres between about 35° and 65° latitude. Westerlies are known to produce cooler weather and precipitation.
See also EASTERLIES.

West European Basin A basin of the Northeast Atlantic Ocean centered at about 45° N and 15° W between Western Europe and the undersea Mid-Atlantic Ridge. The point of greatest depth is roughly 16,200 feet (4,938 m).

west-flowing currents Surface currents generated by the westerlies, occurring in all oceans, in both hemispheres, mostly around the 45° parallel. The direction of a current is dependent on many factors, such as the position of the intervening continental masses relative to wind direction, the influence of the CORIOLIS EFFECT, and changing temperature gradients within the sea.

West Greenland Current A cool, low-salinity current of the North Atlantic Ocean branching from the EAST GREENLAND CURRENT and setting northward along the west coast of Greenland.

West Spitsbergen Current A current of the North Atlantic Ocean setting westward along the east coast of Greenland after sweeping past Spitsbergen. It is a branch of the Norway Current and forms into the EAST GREENLAND CURRENT.

West Wind Drift Eastward-flowing current close to continental Antarctica driven by polar westerlies.

wetlands A low-lying region of marsh or swamp bordering estuaries or other protected, coastal areas that is continually saturated with water and occasionally flooded as a result of spring tides or bulging rivers. Wetlands are regarded as wildlife habitats and in many parts of the world are protected by law.

whale A general name applied to a wide range of marine mammals in the cetacean order, of which there are two major groups—the toothed (odontocetes) and the baleen (balaenoptera) varieties. Whales, which are not fish yet never leave the water, give birth to live young that the mothers nurse with milk. When swimming, whales move their horizontal tails up and down, rather than side to side as fish do. The largest baleen whale is the blue whale, reaching an average length of 100 feet (30 m). The largest toothed whale is the sperm whale of *Moby Dick*, which may grow to more than 70 feet (21 m) long. These warm-blooded animals breathe air and must surface regularly to do so. Whales belong in the order Cetacea, class Mammalia, subphylum Vertebrata, phylum Chordata, kingdom Animalia.

wharf A structure built along a coastline and bordering the edge of a dock, used by vessels for their business on land such as loading and unloading cargo.

Wharton Basin A major basin of the INDIAN OCEAN centered at about 16° S and 98° E. The undersea features of Broken Ridge border it to the south and Ninetyeast Ridge to the west. To the east lies Australia and to the north, Indonesia. Its point of greatest depth is roughly 19,550 feet (5,959 m) near the southern end of the north-south-running Investigator Ridge.

whitecap The white foam that rolls across the top of a wave in unsettled weather. A graybeard is a particularly huge wave with white foam running several feet down its face as the wave breaks during a storm.

White Sea Also called Beloye More, it is a shallow, marginal sea of the Arctic Ocean, running into the northwestern coast of Russia. It is centered at about 42° E along the Arctic Circle.

WHOI *(Woods Hole Oceanographic Institute)* A private, nonprofit research group dedicated to marine science. Fields such as chemistry, marine biology, geology, geophysics, physics, and engineering are pursued at its Cape Cod site, Woods Hole, Massachusetts.
See also Useful websites in the Appendixes.

whorl In a GASTROPOD's shell, the turns and/or twists.

Wilson cycle The hypothesis that an ocean basin has a lifespan with several stages: the opening early stage, which involves crustal uplift and extension with the formation of rift valleys; wide ocean basin development with continental shelves; expansion, in which the system grows unstable; the sinking of the lithosphere to form oceanic trenches; oceanic shrinkage with continental compression and uplift that forms young mountain ranges; then final destruction, in which all oceanic crust is subducted between colliding continents and the tectonic plates became inactive.
See also WILSON, JOHN TUZO.

wind-driven circulation The (typically) horizontal movement of ocean water when driven by wind traveling in a constant direction for a prolonged period.

windward The unsheltered side of an object (island, ship, or other) exposed to oncoming wind. This is the opposite of LEEWARD.

World Ocean Name applied to the water mass covering roughly 71 percent of the Earth. The World Ocean comprises five (formerly four) oceans: the PACIFIC, ATLANTIC, INDIAN, ARCTIC, and the most recently classified SOUTHERN (Antarctic) OCEAN.

GLOSSARY

wharf – World Ocean

worm An invertebrate organism, of which there are thousands of types and four main groups: the flatworms that live in the sea (Platyhelminthes); ribbon worms that live in the sea (Nemetera); roundworms that live in water, on land, or as parasites (Nematoda); and segmented worms that live in the sea and along the shore (Annelida).

wreck Term used to describe the displacement of pelagic-dwelling seabirds due to storm conditions. This can sometimes result in a mass stranding.

xiphactinus More commonly called the bulldog fish, the now extinct xiphactinus was the largest bony fish of the Cretaceous. These predators roamed shallow seas, such as those once covering parts of North America, swallowing their prey whole. However, this tidy dining method may have, for some, served as their demise. Many specimens have been unearthed with the whole, undigested remains of their last meal still within their stomach, as if they had died either during or shortly after swallowing their victim.

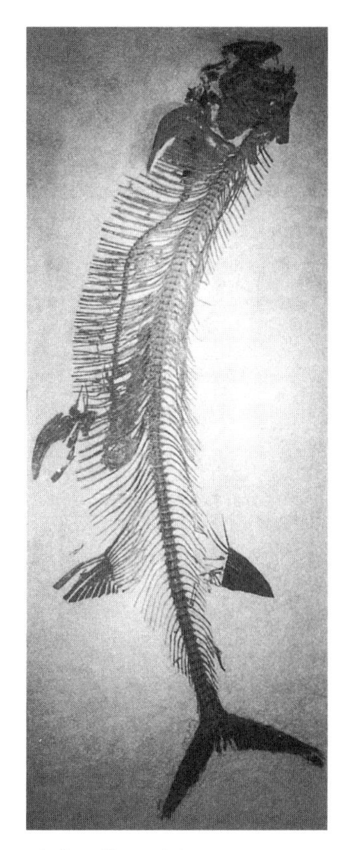

xiphactinus (Photo courtesy: Royal Tyrrell Museum/Alberta Community Development)

yaw A nautical term meaning to swing off course, for instance, as the result of a following sea (wave direction or current running along with the vessel from astern) or a quartering sea (wave direction or current flowing toward the sides of the vessel aft of amidships).

yawl A two-masted sailing vessel in which the mizzenmast is set abaft (behind) of the rudderpost.

Yellow Sea A marginal sea of the eastern Pacific Ocean centered at about 34° N and 122° E along the eastern coast of China. It connects the Bohai Sea, to the northeast, with the East China Sea, to the southeast. Japan and the Koreas lie to the east-northeast, and it connects to the Sea of Japan by way of the Korea Strait.

Yellow Sea Current A warm-water current branched off from the Kuroshio Current that sets northwestward into the Yellow Sea.

Yucatán Current A surface current of the CARIBBEAN SEA setting northwestward through the Yucatán Channel into the eastern portion of the Gulf of Mexico. It is a product of the westward-flowing NORTH EQUATORIAL CURRENT and the GULF STREAM gyre.

zenith The point in the sky that lies directly overhead, as opposed to the NADIR.

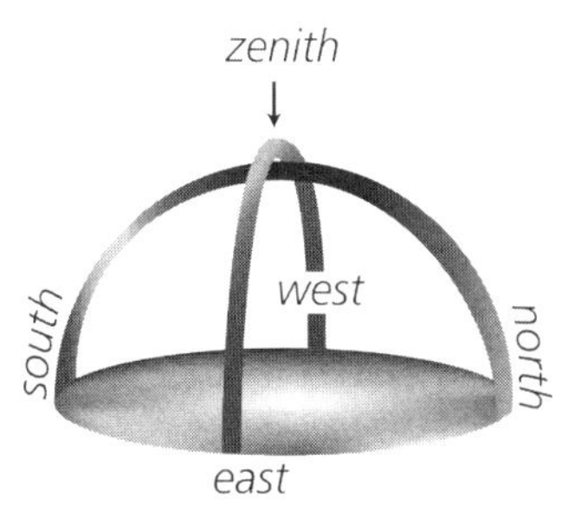

zenith

zonation Parallel bands of distinctive plant and animal associations found within the INTERLITTORAL (intertidal) ZONES and distributed to take advantage of optimal conditions for survival in the physical environment.

zoogeography The distribution of Earth's animals in regard to their geographic location.

zooid An individual member of a colony, such as a bryozoan.

zooplankton Various, chiefly microscopic animal plankton—as opposed to PHYTOPLANKTON, or plant plankton—suspended in the WATER COLUMN. Zooplankton is what many of the ocean's larger inhabitants feed on, consisting mainly of arrowworms, microscopic crustaceans, krill, and forms that do not swim, such as the eggs and larvae of a broad range of animals.

zooxanthellae Photosynthetic forms of algae (usually dinoflagellates) that live symbiotically in the tissue of organisms, such as corals and some MOLLUSKS, and can provide some of the organism's food supply by PHOTOSYNTHESIS.

zygote A cell formed by the union of two gametes, one male one female, by means of sexual reproduction.

SECTION TWO
BIOGRAPHIES

Adhemar, Joseph Alphonse (1797–1862) French mathematician who, in 1842, published the book *Revolutions of the Sea,* in which he incorrectly proposed that because of the equinoxes, the Southern Hemisphere experienced more darkness than daylight per year and the result was the existence of the Antarctic Ice Sheet. He further theorized that the cycle of the equinoxes would cause an ice age in whichever hemisphere was experiencing the longer winter. We now know that, the Antarctic continent itself maintains the ice sheet by keeping the erosive effects of the ocean in check and that the amount of heat received in each hemisphere is the same. Though wrong about some of the details, Adhemar was correct in identifying orbital variations as a possible cause of ice ages, prompting others to explore this field of study.

See also CROLL, JAMES.

Agassiz, Alexander
(Courtesy of NOAA)

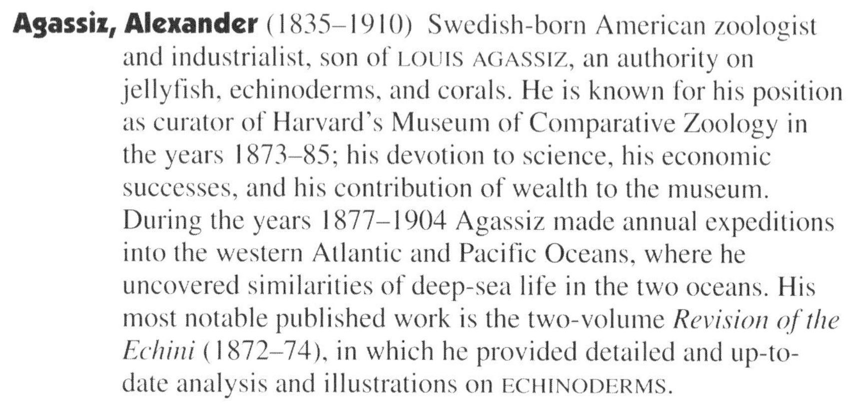

Agassiz, Alexander (1835–1910) Swedish-born American zoologist and industrialist, son of LOUIS AGASSIZ, an authority on jellyfish, echinoderms, and corals. He is known for his position as curator of Harvard's Museum of Comparative Zoology in the years 1873–85; his devotion to science, his economic successes, and his contribution of wealth to the museum. During the years 1877–1904 Agassiz made annual expeditions into the western Atlantic and Pacific Oceans, where he uncovered similarities of deep-sea life in the two oceans. His most notable published work is the two-volume *Revision of the Echini* (1872–74), in which he provided detailed and up-to-date analysis and illustrations on ECHINODERMS.

Agassiz, Elizabeth (Cabot) Carey (1822–1907) American scientist and second wife of the widowed LOUIS AGASSIZ, who assisted in her husband's work and helped establish a marine laboratory on Penikese Island in Buzzard's Bay, Massachusetts. She wrote and published two successful manuals, *Actea: A First Lesson in Natural History* (1859) and *Seaside Studies in Natural History* (1866). Aboard the ship *Hassler,* she voyaged out to sea as assistant to Agassiz and wrote articles on deep-sea dredging. Elizabeth Agassiz also helped in the development of the Natural History Museum at Cambridge, Massachusetts; she founded then become the first president of Radcliffe College. Her writings include *First Lesson in Natural*

History (1859), coauthored by her stepson, ALEXANDER AGASSIZ, and *Seaside Studies in Natural History* (1865). Having outlived her husband, she later wrote *Louis Agassiz, His Life and Correspondence* (1885).

Agassiz, Louis (Jean-Louis Rodolphe Agassiz) (1807–1873) Swiss-American naturalist and geologist devoted to scientific research, known in his time as probably the world's most knowledgeable biologist. Louis had a talent for energizing public interest in science, so great was his love for it. Between the years 1832 and 1846 he acted as professor of natural history at the University of Neuchâtel, where he completed *Recherches sur les poissons fossiles* (Research on fossil fishes), the most complete work in that field to that date. After moving to the United States in 1846, he became professor of zoology and geology at Harvard in 1848; there he later founded the Museum of Comparative Zoology.

Agassiz, Louis (Courtesy of NOAA)

Albert I, prince of Monaco (1848–1922) Monegasque oceanographer and early conservation advocate who founded two oceanographic research institutions, one in Monaco and one in Paris. Each season, Albert-Honoré Charles Grimaldi worked at sea; he released his findings on the GULF STREAM in 1885 and later his findings on Atlantic drift currents. He developed plankton nets and made benthic organism hauls from as deep as 19,000 feet (5,791 m).

Amundsen, Roald (1872–1928) Norwegian explorer known for his numerous Arctic explorations and, most famously, as the man in charge of the first expedition to reach the South Pole (1911). Originally, Amundsen set off to conquer the North Pole; however, as he was making ready to depart, he received news that ROBERT E. PEARY had beaten him to it. Driven by stalwart determination and seasoned experience, he turned on his heel and instead led his team southward into history. His party beat the ill-fated ROBERT F. SCOTT's British party to the South Pole by over a month. Amundsen was also responsible for determining the location of magnetic north and commanded the first single ship to navigate the NORTHWEST PASSAGE (1903–06) and was the first after NILS A. E. NORDENSKJÖLD to traverse the NORTHEAST PASSAGE. Today, there are still many firsts to be won in the field of marine research.

Amundsen, Roald (Archival photograph by Mr. Steve Nicklas/courtesy of NOAA)

Anaximander (c. 611–c. 547 B.C.E.) Greek philosopher who was one of the first minds to offer a natural, rather than mythological, explanation for the process of evolution. Among his beliefs, he stated that animals evolved from a moist environment and human beings from fish.

Andrews, Roy Chapman (1884–1960) Wisconsin-born American explorer, naturalist, and authority on whales who led expeditions for the American Museum of Natural History, New York, and later became its director (1935). Between 1908 and 1914 he voyaged to coasts such as those of Alaska and Asia in order to study marine life. Later he traveled inland, discovering fossil fields in Asia and in the Gobi Desert, unearthing some of the world's oldest known fossils. Among his literary accomplishments are *Whale Hunting with Gun and Camera* (1916), *This Amazing Planet* (1940), and *An Explorer Comes Home* (1947).

Anning, Mary (1799–1847) British geologist, informally educated, yet described as "the greatest fossilist the world ever knew" because of her lifelong association with fossil collecting and the knowledge that arose from it. She is credited as having helped in the 1811 discovery of the first ichthyosaurus (fish lizard) skeleton, the first known to the scientific societies in London.

Archimedes (c. 287–212 B.C.E.) Greek mathematician and inventor who discovered the basic laws of hydrostatics, a branch of physics dealing with fluids. One in particular, the law of buoyancy, called Archimedes' principle, explains why an object seems to lose weight when placed in water. He determined that, when weighed, an object would appear to lose the same amount of weight as the weight of the water it displaced. He correctly concluded that a floating object displaces an amount of fluid equal to its own weight.

Aristotle (384–322 B.C.E.) Greek philosopher and an outstanding natural scientist in his own time, who, among many other achievements, established his personal form of zoological classification, in which animals were more complex than plants. Humans ranked highest on his scale and organisms such as worms and flies lie at the bottom. He named hundreds of organisms, more than 100 of them fish, and the term

Aristotle's lantern to this day is applied to the mouth arrangement of the sea urchin. Around 350 B.C.E. he declared that on maturing, the then very mysterious eel migrated out to sea and disappeared, an explanation that differed from the mythological version, where that eels climbed ashore to mate with snakes and that the skin they shed then became the young eel. Aristotle insisted that only through the method of observation, not philosophy, could the secrets of biological science be revealed, and for years his work was the only arena from which later scientists operated.

Bache, Alexander Dallas (1806–1867) American scientist who, during his long career with the United States Coast Survey, made several contributions to science and helped to establish a lasting interest in marine research. Bache descended from a family of distinguished scientists, including BENJAMIN FRANKLIN, who was his great-grandfather. One of his most notable accomplishments was the work he and his team did charting the Gulf Coast from Texas to Florida. The University of Pennsylvania appointed him professor of natural science, and he also served as an adviser to the United States Navy during the Civil War. In 1836 he was appointed regent of Girard College in Philadelphia, and he was also a regent of the Smithsonian Institution while working with the Coast Survey.

Bacon, Sir Francis (1551–1626) English philosopher responsible for some of the world's greatest literary works. While he lived he actively influenced Europe's scientific community and was acknowledged by the Royal Society of London as an advocate for the advancement of scientific knowledge. Bacon wanted people to look beyond the teaching of the few accepted philosophers, such as ARISTOTLE, and broaden their learning. The title page on his work *Instauratio magna,* which holds his famous *Novum organum* (1620), symbolically depicts a ship sailing out through the Gates of Hercules, which at the time was viewed as the very limit to possible exploration. Other works include *The Proficience and Advancement of Learning* (1605), in which he strives to influence the development of natural and cultural sciences and of modern philosophy, and *The New Atlantis* (1626), an unperfected work in which he describes a journey he made into the "South Sea," embarking for Asia from Peru.

Baffin, William (c. 1584–1622) English navigator who explored the Arctic, looking for the NORTHWEST PASSAGE. With that as his goal, he made a voyage in 1612 as captain of the *Patience* and another in 1615 captaining the *Discovery*. Both attempts were fruitless. During these trips, however, he recorded many highly accurate navigational observations. In 1616 he sailed the *Discovery* as far north as 78° N, a penetration into the Arctic that held the record for over 200 years. Later, between 1617 and 1621, he went to work exploring the Mediterranean, Persian Gulf, and Red Sea regions until his death.

Balboa, Vasco Núñez de (c. 1475–1519) Spanish explorer and conqueror famous not only for leading Spanish conquest in the Western Hemisphere but also for being the first European to sight the Pacific Ocean, in 1513, from a mountaintop in what is modern-day Panama. He named it Mar del Sur, meaning the South Sea; it was later renamed the Pacific by FERDINAND MAGELLAN.

Ballard, Robert D. (Photo by M. Granger)

Ballard, Robert D. (b. 1942) American engineer who launched a new era in deep-sea exploration by inventing the uncrewed submersible, which he called telepresence. Ballard's *Argo-Jason* system is a deep-sea submersible equipped with three highly specialized cameras capable of filming in near darkness. His discovery of the wreck *Titanic* in 1985 won him worldwide recognition for his achievement in this new technology. After his 1985 success, he made several other deep-sea findings that are ranked as the most significant in the 20th century. One such is the discovery of sites on the ocean floor where mineral-rich water is discharged from smoking, geyserlike stacks called HYDROTHERMAL VENTS. To the amazement of marine biologists, at depths thought to be barren of life, large quantities of previously unidentified creatures were found, sparking new debate about the origin of life. Ballard is also responsible for establishing the Jason Project, a program promoting interest in science among students.

Banks, Sir Joseph (1743–1820) British naturalist who accompanied JAMES COOK on his global voyage aboard the *Endeavor*, between 1768 and 1771, and collected biological samples as an unpaid naturalist. He also helped establish botany as an academic discipline. By 1776 he was a member of the science community's prestigious Royal Society of London, and in 1778 he was elected its president.

Barents, Willem (c. 1550–1597) Dutch navigator and seeker of the NORTHWEST PASSAGE. His is known as the first European expedition (1596) to winter so far north, 600 miles (970 km) north of the Arctic Circle, on an island of his discovery, Nova Zemlya, after becoming icebound. He did not succeed in finding the Northwest Passage, but he explored Barents Island, Spitsbergen, and Nova Zemlya island in the Arctic Ocean. In 1597, which was the year of his third voyage, he died while returning from the Arctic in an open boat. The BARENTS SEA honors his name.

Barrell, Joseph (1869–1919) American geologist who declared that sedimentary rock was produced by the action of wind, rivers, and ice, not merely formed under oceans. He is the person who coined the terms *lithosphere* and *asthenosphere* and was the first geologist who fully realized the potential of the new technology radioactive dating. In the year of his death, 1919, Barrell was elected a member of the American National Academy of Sciences.

Barrow, Sir John (1764–1848) English geographer and second secretary of the Admiralty from 1804 to 1845 who promoted the exploration of the Arctic. He encouraged the voyages of SIR WILLIAM E. PARRY, SIR JOHN FRANKLIN, and SIR JOHN ROSS with the goal of furthering the world's existing knowledge of navigation and the area's geographic features. He wrote *Voyages of Discovery and Research in the Arctic Regions* (1846), as well as accounts of his travels in China. Cape Barrow, Barrow Strait, and Point Barrow all honor his name.

Bartlett, Robert Abram (1875–1946) American Arctic explorer born in Canada, famous for his navigational skills through iceberg-ridden waters. Between 1905 and 1909 he piloted the *Roosevelt* for ROBERT E. PEARY's expeditions. During an expedition in 1913, his ship *Karluk* became icebound near Wrangel Island and was eventually crushed. On foot he traveled over ice to Siberia and returned with rescuers for his crew. Later, he made 20 Arctic voyages aboard the *Morrissey.* His writings include *The Last Voyage of the Karluk* (1916) and *Sails over Ice* (1934).

Barton, Otis (1899–1992) American engineer and oceanic explorer who designed the first bathysphere (a large, spherical steel diving chamber suspended on a cable) for the use of the American naturalist CHARLES W. BEEBE, with whom he began an association in the 1920s. In 1930 he and Beebe made the first deep-sea dive to a depth of 1,425 feet (434 m), and in 1934 to a record depth of 3,028 feet (923 m). Barton donated the bathysphere to the New York Zoological Society; for this gift he was awarded a lifetime membership in the class called "Associate Founder." In 1939 he and Beebe had a falling-out over Barton's movie *Titans of the Deep,* which they never reconciled because Beebe took credit for it. Barton never carried out another project for the Society.

See also BATHYSCAPHE.

Bass, George (1771–1803) English explorer and naval surgeon known as the man who, in 1798, proved the existence of an ocean passage lying between Tasmania and Australia that connects the Indian with the Pacific Ocean. This passage was eventually named the Bass Strait. Before that event, however, Bass, together with MATTHEW FLINDERS, served aboard the HMS *Reliance,* and later the two explored Botany Bay in an eight-foot (2.4-m) boat, the *Tom Thumb.* In 1799 Bass was made a member of the Linnean Society of London, in recognition of his achievements in the scientific study of southern Australia's flora and fauna. In 1803, while traveling from Australia to South America, the cargo ship on which he was sailing disappeared at sea.

Baudin, Nicholas (1754–1803) French explorer and navigator who, beginning in the late 1700s, commanded several scientific voyages into waters of the Caribbean, eastern Asia, and South America. In 1800 he embarked on a French-sponsored expedition to explore and, it was hoped, established a French claim to Tasmania (then called Van Diemen's Land). Baudin is named as the person who succeeded in mapping the island; however, at one point during his voyage he encountered the expedition of the British navigator MATTHEW FLINDERS, who turned over to Baudin his newly charted maps of Australia. Baudin died of illness in 1803, before the party's return, and the shipboard scientist FRANÇOIS PERON and Captain Louis de Freycinet, who took command of the mission after Baudin's death, recorded an account of the voyage. On their return home Baudin's expedition was given undue credit for Flinders's work.

Beaufort, Sir Francis (1774–1857) Irish-born British admiral and hydrographer (studier of the seas) most famous for his invention (1805), the BEAUFORT WIND SCALE, an instrument designed to benefit mariners (and meteorologists) as it would indicate wind velocity and measure its effects on a vessel under full sail. In 1787, he joined the British navy, in which he served for 20 years; he was elected to the Royal Society of London in 1814 and in 1832 was assigned as official hydrographer to the British Admiralty.

Beebe, Charles William (1877–1962) American explorer, writer, and naturalist who made contributions to marine science in several different ways. In 1899 he was appointed curator of ornithology (the study of birds) for the New York Zoological Society; in 1916 he became their director of tropical research. Beebe led many biological research expeditions to lands such as China, Japan, the Himalayas, and Borneo, and throughout his life he wrote nearly 300 books and articles. He is probably most known, however, for making a record-breaking deep-sea dive in a tethered bathysphere, accompanied by his fellow designer, the American engineer OTIS BARTON, down Bermuda's Puerto Rico Trench in 1934. They reached 3,028 feet (923 m).

See also BATHYSCAPHE.

Belcher, Edward (1799–1877) Canadian-born British navy admiral who surveyed many coastal areas throughout his career. In 1825 he surveyed the western coast of Canada, including the Bering Strait; between 1830 and 1833 he surveyed the western and northern coasts of Africa; between 1836 and 1846 he surveyed the western coasts of North and South America and the coasts of China, Borneo, and the Philippines. In 1852 he led a rescue expedition to find the missing explorer SIR JOHN

FRANKLIN but lost his ships to the ice. Later, he wrote a book about the failed trip, *The Last of the Arctic Voyages* (1855). He was knighted in 1867 and then raised to admiral in 1872.

Bellingshausen, Fabian Gottlieb von (1778–1852) Russian naval officer and explorer best remembered as the leader of the first expedition to circumnavigate Antarctica, in the years 1819–21. With his two ships, the *Mirny* and the *Vostok,* he explored the region south of the 70th parallel, establishing that the South Sandwich Islands, which he named after Tsars Peter I and Alexander I, were not part of the main continent, as previously assumed, but were smaller separate landmasses. Later, during the Russian war with Turkey (1828–29), he was promoted to admiral and subsequently appointed governor of Kronshtadt. The Bellingshausen Sea as well as one of the South Sandwich Islands honor his name.

Belon, Pierre (1517–1564) French naturalist known for his writings on the natural history of fish and birds. Belon began life as an apothecary but later studied with the renowned German physician and professor of botany Valerius Cordus. Later he developed a medical practice in the royal court society of France. Indications point to his being the first to note the anatomical similarities in the vertebral bones of fish and mammals. His writings include *On Aquatic Life* (1553) and the *Natural History of Strange Marine Fish* (1551), which dealt with classifying fish and cetaceans; in it he recognized the latter as being milk-producing air-breathers yet nevertheless classified them as fish. He was murdered on the streets of Paris.

Bering, Vitus (1680–1741) A Dane commissioned by the Russian tsar Peter the Great to determine whether or not a land bridge existed between Siberia and North America. In 1724 he found not land but a passage of water, proving that Asia and America were separate. The waterway was later named the Bering Strait. In 1741, still in the service of Russia, he embarked on the Great Northern Expedition; however, quite soon his ship was wrecked on a tiny island in the western Aleutian chain of Alaska. A few months later Bering died of exposure on the same island, later named Bering Island in his memory; some crewmen survived to pass on the news and information gleaned from the voyage.

See also STELLER, GEORGE WILHELM.

Bernoulli, Daniel (1700–1782) Dutch-born Swiss scientist and authority on physics. In his book *Hydrodynamica,* he established the basic principles of movement within fluid, showing that pressure decreases with an increase in velocity. He also worked on magnetism, ocean tides and currents, and the movements of ships, gaining many awards from Paris's Academy of Sciences. Bernoulli lectured on physics in Saint Petersburg, Russia, and was a member of the Russian Academy of Sciences.

Bigelow, Henry Bryant (1879–1967) American marine biologist and oceanographer notable for his extensive studies of the Gulf of Maine, which were supported both by the U.S. Bureau of Fisheries and by Harvard's Museum of Comparative Zoology. In 1901 he took part in an expedition of ALEXANDER AGASSIZ's to the Maldive Islands; again with Agassiz in 1905 he voyaged into the eastern Pacific Ocean aboard the research vessel *Albatross.* Bigelow was wealthy and well-traveled and became friends with many of the most notable scientists in the blossoming field of marine biology. During his 12-year study of the Gulf of Maine, Bigelow set modern standards in marine science. His findings concerning the Gulf of Maine are still referenced today.

Blagden, Sir Charles (1748–1820) English physician and chemist who in 1788 proposed Blagden's law, a scientific principle stating that the lowering of a solution's freezing point was proportional to the concentration of the solute present; for example, seawater, meaning salt water, freezes at a lower temperature than freshwater. Blagden graduated from the University of Edinburgh, Scotland, in 1768 and served as medical officer to the British army until 1814, working during that time with such eminent scientists as the British physicist Henry Cavendish, the Scottish inventor James Watt, and the French chemists Antoine Lavoisier and Claude Berthollet.

Bligh, Captain William (1754–1817) English naval officer who sailed under the leadership of the English navigator JAMES COOK between the years 1772 and 1775 during Cook's second voyage. He is most famous as captain of the *Bounty,* a ship whose crew mutinied (1789) and set Bligh adrift in a boat with 18 men who remained loyal to him. Only through skill of seamanship did he and his men survive. Over the course of 17 days the castaways traveled 3,618 miles (5,823 km), suffering many hardships along the way, finally to land on the island of Timor near Java.

Booth, Felix (1775–1850) British gin merchant and alderman (official just below a mayor in rank) of the city of London who financed his friend SIR JOHN ROSS's 1829–33 expedition to search for the NORTHWEST PASSAGE, when the British Admiralty would not. He gave Ross £18,000 toward the cost of the mission, which Ross used to purchase a ship, the *Victory*. The Gulf of Boothia and Boothia Peninsula (the northernmost tip of North America) were named by Ross in recognition of Booth's generous financial participation.

Borchgrevink, Carsten (1864–1934) Norwegian-Australian explorer who participated in the first expedition to set foot on the Antarctic continent in 1894 while on a whaling expedition led by H. J. Bull, and who, while in command of the British-owned ship the *Southern Cross*, went in search of the magnetic South Pole in 1898. His party was the first to winter on the Antarctic continent, where they made magnetic observations and collected plant and animal specimens. In 1899 more magnetic observations were made around Franklin Island and Borchgrevink's team determined the south magnetic pole must be much farther north and west than previously assumed.

Bougainville, Louis-Antoine de (1729–1811) French navigator and naval officer who made notable contributions to scientific fields and was a member of the Royal Society of London by 1754. He is known to be the first Frenchman to have sailed around the world (1766–69), in search for new lands. Accompanying Bougainville on this voyage were astronomers and naturalists, who together made many discoveries in science and geography. During the trip he rediscovered the Solomon Islands, the largest of which, Bougainville, Papua New Guinea, is named for him. He also sailed into Tahiti, Samoa, and the Hebrides Islands; sighted the Great Barrier Reef, then sailed north toward New Britain. He spent time exploring the Moluccas and Jakarta then headed back for France. Bougainville recorded his journeys by publishing the two-volume *Account of a Voyage around the World* in 1771–72.

Bourne, William (c. 1535–1582) British innkeeper who wrote the book *Treasure for Travelers* (1578), in which he described the possibility of ballast tanks on a submarine, enabling it to rise and descend, though he never actually built them himself. In it he also described his concept of the flow of the Atlantic Ocean and the little-understood power of the Moon on the tides and currents.

Bowditch, Nathaniel (1773–1838) American mathematician, navigator, and astronomer who wrote a book on the principles of navigation and their methods of application. The book, *The New American Practical Navigator* (1802), is still in use today by the U.S. Navy, being continually revised, under the title *The American Practical Navigator.*

Boyle, Robert (1627–1691) English natural philosopher remembered for his instrumental role in the founding of the Royal Society of London and as one of the founders of modern chemistry. He invented a vacuum pump and used it in formulating what is known as Boyle's law, a law of physics explaining how the pressure and volume of a gas are related. He also made studies of the propagation of sound.

Brendan, Saint (c. 485–c. 583) Irish monk who is rumored to have been the first European to travel the Atlantic Ocean. To some he is known as Brendan the Navigator, a monk who yearned to travel. To satisfy his nomadic nature, Brendan organized and then embarked on long journeys, accompanied sometimes by large parties of people. In his literary work *Navigatio sancti brendani abbatis,* Brendan tells about islands that could have been the Canaries, the Azores, Iceland, or the Faroe and the Shetland Islands. His boats were constructed of caulked hides stretched over wicker frames.

Britton, Elizabeth Gertrude Knight (1858–1934) American botanist and early conservationist who specialized in bryology, the study of the mosses. Between 1881 and 1930 she published an impressive 346 papers, in publications put out by societies such as the *Bulletin of the Torrey Botanical Club* and the

Sullivant Moss Society, in which she served as founder and president (1916–19). In 1949 that society became the American Bryological Society. Britton was unofficially in charge of the moss collections at Columbia College, New York; had 15 species and one moss genus *(Bryobrittonia)* named for her; and was instrumental in the establishment of the New York Botanical Gardens.

Buchanan, John (1844–1925) British oceanographer and chemist who between 1872 and 1876 took part in the *Challenger* mission, designed a water sampling bottle that he used during the voyage, and provided one of the first reliable charts showing the surface temperatures of the ocean. During 1885–86, while sailing his own vessel off the west coast of Africa, he discovered the existence of the Equatorial Undercurrent and also showed how vertical currents carried colder water to the ocean surface.

Buffon, George-Louis Leclerc, comte de (1707–1788) French naturalist best known for his famous 36-volume work *Histoire naturelle, générale et particulière* (General natural history, 1749–89). This was one of the earliest accounts of biological and geological history not linked with the Bible. In it he provided complete and naturalistic descriptions of zoological, mineralogical, and botanical productions of the Earth based on scientific observation and experimentation. In 1753 he became a member of the French Academy of Sciences.

Cabeza de Vaca, Álvar Núñez (c. 1490–c. 1557) Spanish explorer who, in 1527 in the role of treasurer, embarked on an expedition to colonize Florida. After their shipwreck on an island near Texas, Álvar Núñez, as he was known, and a few other survivors experienced great hardships. They were enslaved by Native Americans until finally Núñez led an escape overland and eventually returned to Spain.

Cabot, John (c. 1450–1499) English navigator and explorer who attempted to find a shorter western route to Asia than the one recently discovered by CHRISTOPHER COLUMBUS. In May 1497, backed by King Henry VII of England, the Genoa-born Cabot (Giovanni Caboto in Italian) set to sea in the *Matthew* with a crew of 18 men. In June he spotted land, which he thought to be Asia but was in fact either Newfoundland or Cape Breton Island. Cabot claimed the land for England. Sailing the coasts, he did not find the hoped-for riches of spice and jewels, but he did discover the abundant fishing grounds of the Grand Banks, which he reported after his return to England. The king opened the region for fishing, still thinking it was the Orient. In 1498 Cabot set out on a second voyage to the same area, this time with five ships and 300 men. Storms claimed all ships but one, and Cabot and his party were probably lost at sea. However, by some accounts he either stayed in the new land or returned to England for supplies and never set sail again.

Cabot, Sebastian (c. 1476–1557) English navigator and mapmaker who was the second son of JOHN CABOT. In 1508, commanding two English ships, Cabot set sail in search of the NORTHWEST PASSAGE, a route to Asia through the Arctic Ocean. His crew refused to travel farther west than what is now Hudson Bay, and Cabot explored only the eastern coast of North America. In 1527 he again set sail, this time from Spain, with four ships, seeking the Spice Islands. He instead explored the eastern coast of South America and returned to Spain in 1530. In 1551 he founded the Muscovy Company of Merchant Adventurers, an English trading organization that financed expeditions to search for the Northwest Passage.

Cabral, Pedro Álvarez (c. 1467–c. 1528) Portuguese navigator who voyaged, by chance, to the land of Brazil, which he claimed for Portugal. In doing so, he helped develop for his homeland a large overseas empire in the 1500s. Originally, Cabral had been assigned to take over the work of VASCO DA GAMA (still living at the time), and in 1500 he sailed from Belem, near Lisbon, with 13 ships and a crew of 1,000, planning to follow da Gama's route to India by rounding the Cape of Good Hope, in Africa. They were carried off course and ultimately landed on the eastern shores of Brazil.

Cabrillo, Juan Rodriguez (d. 1543) Portuguese explorer and soldier for Spain who, under the orders of Antonio de Mendoza, Spanish ruler of Mexico, led the first European expedition to what is now the California coast. In 1542 Cabrillo sailed from Navidad, Mexico, and reached San Diego Bay within three months. He continued northward, exploring Santa

Catalina Island, Santa Barbara Channel, Monterey Bay, and San Miguel Island. He again turned southward after weathering a storm in the safety of San Francisco Bay. Cabrillo died on San Miguel Island.

Calvin, Melvin (1911–1997) American chemist notable for his work in PHOTOSYNTHESIS, which won him a Nobel Prize in chemistry in 1961. He traced the process of a plant's converting carbon dioxide and water into sugar by using radioactive carbon-14 to follow the movement of carbon through sequences of chemical reactions. He is also known for his work with certain species of plants that produce fuel oil.

Cano, Sebastian del (c. 1476–1526) Spanish navigator and captain of the *Concepción,* one of the five vessels under the direction of FERDINAND MAGELLAN during his expedition to circumnavigate the globe. After the death of Magellan in April 1521, del Cano took over as leader. Lacking adequate crew, the *Concepción,* being the least seaworthy of their three remaining ships, was scuttled. Del Cano finished the voyage begun by Magellan and on returning to Spain on board the *Victoria* in the year 1522 received all the credit for the mission.

Carpenter, William Benjamin (1813–1885) English biologist and professor of physiology at the Royal Institution, London, notable for his valuable research on the unicellular organism foraminifera. He also took part in deep-sea exploration between 1868 and 1871.

Carson, Rachel (1907–1964) American marine biologist and writer of numerous books on the sea and coasts. She spent most of her career (1936–52) as an aquatic biologist at the U.S. Fish and Wildlife Service. Through her literary works, which received awards for their scientific accuracy, she voiced concern over the use of pesticides, warning that they were contaminating the food sources of wildlife and potentially those of humans. Her arguments led to restrictions of chemical pesticide uses in many regions of the globe.

Cartier, Jacques (1491–1557) French navigator who in 1534, while searching for the NORTHWEST PASSAGE, discovered the Saint Lawrence River in Canada. Twice more he sailed to the

Canadian coast, in 1536 and 1541, and his early domination of the area laid the groundwork for later French claims.

Castro, João de (1500–1548) Portuguese navigator who pioneered the first technique for the mapping of magnetic north based on observations of the Sun. Between the years 1538 and 1541, de Castro made highly accurate measurements of deviation in his ships' logs, which gave rise in subsequent years to the regular practice of navigating with a compass. Mapping the position of magnetic north was very important to mariners since its location varies with one's position on Earth.

Cavendish, Thomas (1560–1592) English explorer known as the second Englishman to sail around the world. His first exploration in 1585 was an attempt to colonize Virginia; however, his greatest interest, it seemed, was not in colonization, but in piratical attacks on Spanish shipping. His second voyage in 1586 took him and his party toward the Straits of Magellan. Along the way, they engaged in many skirmishes, which earned Cavendish fame and riches on his return to England in 1588. Cavendish lóved to squander money and was therefore always looking for new and interesting ways in which to obtain it. In 1592 he again set sail to look for a trade route to the Orient; however, his ships encountered storms, icing conditions, and hardship. The party turned back to England. Cavendish died on the voyage.

Champlain, Samuel de (1567–1635) French-born governor of Canada known as the father of New France in North America. In 1603 he voyaged to Canada as royal geographer on a fur trading expedition. From 1604 to 1607 he explored Canada's east coast, and in 1608 he founded a trading post that eventually became Quebec; he remained there as governor until his death. Lake Champlain is named in his honor.

Chancellor, Richard (d. 1556) English navigator who in 1553 served as second in command on SIR HUGH WILLOUGHBY's voyage in search of the NORTHWEST PASSAGE. A storm separated their ships and Chancellor found his way to the White Sea alone, commanding the vessel *Edward Bonaventure*. He reached the site of modern Arkhangel'sk, Russia, and traveled overland to Moscow to secure a treaty that would allow English ships freedom to trade. In 1555 he again sailed to Moscow, but on the return voyage, his ship succumbed to the forces of the sea.

Chappe d'Auteroch, Jean-Baptiste (1728–1769) French astronomer and student of marine science who, despite reports of contagious diseases running amok throughout the region, set sail for Baja California to observe the transit of Venus. He took with him a hydrometer and throughout the voyage observed the specific gravity of the sea. On two occasions he lowered a thermometer wrapped in canvas to a depth of 100 fathoms (600 feet [183 m]) to compare the temperatures below with those at the surface. On arrival at his destination, he successfully observed the transit of Venus then decided to stay on an extra couple of weeks to observe a lunar eclipse, during which time he contracted yellow fever. The expedition cost him his life. A book published after his death, *The 1769 Transit of Venus, The Baja California Observations of Jean-Baptiste Chappe d'Auteroche, Vicente de Doz, and Joaquín Velázques Cárdenas de León*, describes his observations.

Chichester, Sir Francis Charles (1901–1972) British pioneer air navigator and yacht sailor who began his days as an aviator, but during the world's first long-distance solo flight in a seaplane (1931), which he flew from Australia to Japan, he crashed and was badly injured. It was not long afterward that he turned to sailing. From 1940 to 1945 he wrote navigation instruction manuals for the Air Ministry and in 1945 he began the Francis Chichester Map and Guide publishing house, in which he served as secretary, publisher, sales representative, and map designer, creating tens of thousands of wartime Air Ministry jigsaw puzzle maps that he sold to London retailers. In 1960, aboard *Gipsy Moth III*, he won the first solo transatlantic yacht race, and in 1967 he completed the first solo circumnavigation of the world in his yacht *Gipsy Moth IV*, taking 107 days to sail from Plymouth to Sydney and 119 days for the return trip. He was knighted in Greenwich, England, on July 7, 1967.

Clapp, Cornelia M. (1849–1934) American zoologist and talented student of LOUIS AGASSIZ, known as a pioneer of zoological study within the laboratory. After receiving a variety of degrees, she became a professor at Mount Holyoke College and also, in 1888, began active research at the newly established Woods Hole Marine Biology Laboratory. Standing since 1924 at Mount Holyoke College in South Hadley, Massachusetts, is the Clapp Laboratory, named for her, a member of the class of 1871.

Columbus, Christopher (1451–1506) Italian-Spanish navigator and one of the greatest navigators of all time. He navigated his ships by DEAD RECKONING, knowing just enough celestial navigation to measure latitude from Polaris, the North Star. Between 1492 and 1504 he made four major voyages to explore new lands, but he is most famous for his first, the discovery of the Americas in the year 1492. On the basis of the incorrect belief that the world was smaller that it is, Columbus sailed into the uncharted west seeking a shorter trade route to Asia. Queen Isabella of Spain agreed to finance his voyage, persuaded by his promise of a wealth in trade goods on his return. Embarking with three ships, the *Santa Maria* and the smaller *Pinta* and *Niña,* he ultimately landed on the beach at Fernandez Bay, San Salvador, Bahamas, and insisted it was an island of the Indies; as a consequence, Columbus's party wrongfully regarded the gentle Arawaks living there as "Indians."

Cook, Captain James (1728–1779) English navigator who was the first to use technology—the JOHN HARRISON chronometer in 1735—together with astronomy to navigate. Acting as lieutenant in command of *Endeavor,* he set sail in 1768 on his first major voyage to the island of Tahiti, carrying British astronomers with orders to observe a transit of the Sun by the planet Venus. However, private instructions were issued for him to search for the "unknown southern land," Antarctica, as well. Cook found no southern continent and its existence remained legend. In 1769 he took partial possession of both New Zealand islands. In 1770 he discovered what is now known as Australia, charted it, and claimed it for Great Britain under the name of New South Wales. Later, as commander of *Resolution,* which took to sea in 1772 accompanied by the *Adventure,* he successfully circumnavigated Antarctica, but because of bergs and floes, the actual sighting of land eluded him. Yet he proved once and for all that Antarctica, or Terra

Australis, was not part of Africa. In 1773 the captain made the first crossing of the Antarctic Circle and discovered the islands named for him, the Cook Islands. What is more, in 1774, he discovered several Pacific islands, including New Caledonia and Niue, and charted the New Hebrides, Easter Island, and the Marquesas. In 1775, Cook, born the son of a farmer, was awarded the Copley medal for scientific achievement and was made a fellow in the Royal Society of London. In 1778 he became the first known European to discover the Hawaiian Islands, naming them the Sandwich Islands, for the earl of Sandwich, Britain's chief naval minister. A year later, after returning to the islands from a vain attempt to discover a NORTHWEST PASSAGE along the North American continent, he was killed—stabbed to death in a fight with islanders over the theft of a skiff.

Coriolis, Gaspard-Gustave (1792–1843) French mathematician and engineer responsible, for, among other accomplishments, describing the inertial force called the CORIOLIS EFFECT. This theory proved to be very significant in the study of meteorology, ballistics, and oceanography. Coriolis is also credited with coining the term *kinetic energy*.

Cortés, Hernán (1485–1547) Spanish explorer and conqueror of Mexico who is remembered as the first European to sight California (1536). In 1519 Cortés set sail from Cuba toward the Yucatán Peninsula, Mexico, and proceeded to conquer Mexico City. He forayed through Central America and emerged on the Pacific Coast, where he had two crude ships built, with which he used to explore the Baja region. In 1540 he returned to Spain.

Cousteau, Jacques-Yves (1910–1997) French naval officer, marine explorer, author, and documentary filmmaker who in 1943, along with his colleague, Émile Gagnan, was responsible for designing the Aqua-Lung, or scuba tank, which allowed divers to remain underwater for hours. Through his tireless use of submersible and motion picture technology, Cousteau played a major part in popularizing interest in the marine environment during the mid-20th century. He had a passion for the sea and wished to educate the world to understand the fragility of the marine environment. Aboard his ship, *Calypso,* he made numerous underwater films, from full-length motion pictures to shorts, including two that each earned an Academy Award as best documentary, *The Silent World* (1956) and *World Without Sun* (1966). It would be quite surprising to find anyone who had any interest in the sea unfamiliar with the name Jacques Cousteau.

Croll, James (1821–1890) British geologist who theorized that ice ages are caused by eccentricities in the Earth's orbit. Ice ages cause global drops in sea levels as water that is normally liquid is taken up into the expanding glacial regions and trapped as ice. In 1859 he was asked on as caretaker at Anderson's College, Glasgow, and in 1867 was made resident geologist at the office of the Geological Survey, Edinburgh, where he was keeper of the maps and correspondence until 1880. He wrote *The Philosophy of Theism* (1857); *The Philosophic Basis of Evolution* (1890), which deals with physical geology, such as ocean currents and climate; and *Climate and Time* (1875), which addresses possible causes of the glacial epoch.

Cuvier, Georges-Léopold-Chrétien Frédéric Dagobert, Baron (1769–1832) French anatomist who almost single-handedly founded the scientific discipline of paleontology. It has been said that his was one of the sharpest minds in the history of science. Cuvier established the fact that life-forms that once existed were now extinct and was responsible for classifying animals into four branches: Vertebrata (vertebrates), Articulata (arthropods and segmented worms), Mollusca (meaning at that time, all soft-bodied invertebrates of bilateral symmetry), and Radiata (coelenterates and echinoderms). Among his most successful published works are the *Tableaux elementaire de l'histoire naturelle des animaux* (1797) and *Leçons d'anatomie comparée* (1800–05). He spent most of his career at the Museum of Natural History in Paris.

Dampier, William (1652–1715) English buccaneer, navigator, and surveyor who circumnavigated the world (1683–92), an accomplishment that gave him extensive knowledge of hydrography and exploration. Between these adventurous years he rode the high seas of the South Pacific, pirating, and investigated New Holland (Australia), China, and New Guinea. He kept a meticulous log, made charts and conducted surveys, and then published his accounts in a journal, *A New Voyage around the World* (1697). This work helped stoke the fires of curiosity about the Pacific region, and in 1699 the British Admiralty financed another voyage by Dampier to what he described as "barbaric" Australia.

Darwin, Charles Robert (1809–1882) British scientist and naturalist who revolutionized biological thinking by challenging the Christian belief of divine creation with his theory of the evolution of organisms through natural selection. In 1831, Darwin, a graduate of Cambridge University, was taken aboard the survey ship HMS *Beagle*, captained by ROBERT FITZROY, as an unpaid naturalist on a scientific voyage around the world. During this time, Darwin became fascinated with the study of coral reefs; on his return he published *The Structure and Distribution of Coral Reefs* (1843), which detailed the formation of coral atolls. His observations aboard the *Beagle* were fundamental to the development of his theories, and in 1859 he published *On the Origin of Species*, describing his theory of evolution.

Darwin, Erasmus (1731–1802) English physiologist and poet, grandfather of CHARLES ROBERT DARWIN and Francis Galton (a British anthropologist). Erasmus was a freethinker whose extensive, expressive works in prose centered on natural history, evolution, botany, and taxonomy. His writings include the intuitive *Zoonomia* (1794–96), which anticipated later theory, and *The Botanic Garden* (1789–92).

Darwin, Sir George Howard (1845–1912) English geophysicist and mathematician who posed the controversial theory that in ancient times the Moon was much nearer the Earth and that its effects on the tides caused the rotation of the Earth to slow, which then caused the Moon to recede.

Davis, John (1543–1605) English navigator and Arctic explorer who, beginning in 1585, embarked on three back-to-back voyages in search of the NORTHWEST PASSAGE. On his last attempt he made it as far as latitude 73° N, where he discovered a new strait, which was named the Davis Strait. In 1592, with his fellow navigator THOMAS CAVENDISH, Davis discovered the Falkland Islands. Davis improved methods of seamanship by completing two books on navigation and developing a navigational aid called the backstaff, which the French called the "English quadrant" and the English called the "Davis quadrant".

Deacon, Sir George (1906–1984) British oceanographer who was influential in the furthering of marine science in Britain through his innovations in methods to measure deep-sea currents and surface waves. He conducted extensive research in the waters off Antarctica and was a member of the Royal Society of London.

Deitz, Robert Sinclair (1914–1995) American marine geologist, lieutenant colonel, and self-proclaimed astrogeologist who studied under FRANCIS P. SHEPARD and was the first, along with his fellow graduate student K. O. Emery, to describe submarine phosphorites off the coast of California. Deitz made significant contributions in the fields of marine geomorphology, geology, and oceanography. After a stint in the military (1941–50), Dietz traveled extensively, participating in several oceanographical expeditions, mainly in the Pacific Ocean. He was an early advocate of the continental drift theory, coined the terms *seafloor spreading* and *astrobleme* (describing the geographic formation that follows the impact of a meteorite), named the Emperor Seamounts, and spoke of a force that was pushing them northward. He was associated with SCRIPPS INSTITUTION OF OCEANOGRAPHY, NOAA (the National Oceanic and Atmospheric Administration), the Naval Electronics Laboratory, the Office of Naval Research, and the Japanese Hydrographic Office and was a visiting professor at several universities. He was the recipient of several honors over his brilliant career. Among his works are coauthorship with Jacques Piccard of *Seven Miles Down: The Story of the Bathyscaphe Trieste* (1961) and collaborations on the books *Submarine Geology* (Shepard, 1948) and *Submarine Canyons and Other Sea Valleys* (Shepard and Dill, 1966).

Deryugin, Konstantin Mikhailovich (1878–1938) Russian oceanographer and expert on the White Sea. Deryugin was a dedicated man of many hats. Aside from being a professor in Leningrad and a pioneer in the methodology of oceanography, he was manager of the Oceanic Division of the State Hydrological Institute, Leningrad; he conducted major research on classifying aquatic arctic biota; he reestablished the Kola meridian water-sampling station; and he helped organize over 50 expeditions that introduced to science the "new world of organisms."

Dezhnyov (Dezhnev), Semyon Ivanovich (1605–1672) Russian explorer who in 1648 was the first to discover the Bering Strait, but his findings went unnoticed. Later, VITUS BERING rediscovered it in the early 1700s. Cape Dezhnyov, Russia, is named for him.

Dias, Bartolomeu (c. 1450–1500) Portuguese navigator known as the first to sail around the Cape of Good Hope, Africa, in 1488. In 1487 King John II of Portugal commissioned Dias to take a flotilla of three ships southward along the African coast to try to locate its terminus in the hope that a water route to Asia could be found. Dias's discovery indeed helped establish travel between Asia and Europe. In 1497 he captained one of the ships accompanying VASCO DA GAMA on a trip to India, and in 1500 he took part in a 13-vessel expedition led by PEDRO ÁLVAREZ CABRAL, also heading for India. Dias's ship strayed off course and sank on its return from Brazil.

Dohrn, Felix Anton (1840–1909) German zoologist responsible for founding the world's first marine biological research station, Stazione Zoologica, in Naples, Italy, in 1873. Dohrn was a friend of THOMAS H. HUXLEY and studied under ERNST HAECKEL. He was a major figure in early arthropod phylogenetics (evolutionary history of an organism) and promoted the annelid theory of vertebrate origins.

Donati, Vitaliano (1713–1763) Italian biologist and professor of botany at the University of Turin. He is known to have greatly improved the university's Botanical Gardens. Donati's book *The Marine Natural History of the Adriatic* (1750) encompasses a wide scope of information on the hydrologic and biologic characteristics of the Adriatic Sea.

Drake, Sir Francis (c. 1540–1596) English navigator and ambitious military commander who became very popular in 16th-century

England. In 1567, Drake set sail for the West Indies in command of the *Judith;* on arrival his party suffered losses at the hands of the Spaniards. He returned several times afterward to exact revenge, activities that made him famous in England. Between the years 1577 and 1580 he became the first English explorer to sail around the world. Drake embarked into the Atlantic Ocean with five ships, heading toward the Pacific Ocean. At the Straits of Magellan, South America, bad weather pummeled all but the ship *Golden Hind,* on which he crossed the Pacific, rounded Africa by the Cape of Good Hope, and returned to England. In 1585 he sailed with 25 ships to the Spanish Indies and returned to England with a large cargo of goods. By now his seamanship and bravery under fierce encounters had earned him even more honor. In 1595 he sailed once again to the West Indies, where he died of dysentery.

Driver, William (1803–1886) American sea captain famous as the man who first labeled the American flag "Old Glory." He flew the flag, which was a gift, aboard the *Charles Doggett* on his voyages around the world. The same flag now is on display in the Smithsonian Institution.

Dumont d'Urville, Jules-Sébastien-César (1790–1842) French naval officer who explored the South Pacific Ocean and Antarctica. Between 1822 and 1840 Dumont d'Urville made three circumnavigations, during which he investigated many lands, including Australia, New Guinea, and the coast of Antarctica, which he called Adélie Coast. The D'Urville Sea (off Adélie Coast) and Cape d'Urville, Irian Jaya, Indonesia, were named for him, as was D'Urville Island (off New Zealand). On his voyages he also conducted deep-sea observations; however, because of the crude thermometer technology of the age he wrongly concluded that waters below 500 fathoms were at a constant 40° F (4.4° C). Dumont d'Urville applied the same wrong reading of 40° F (4.4° C) to the waters that range from the surface on down between latitudes 40° S and 60° S.

Dutton, Clarence Edward (1841–1912) American geologist who concluded that oceanic rock was heavier than that which made up the continents. Between 1875 and 1890, Dutton worked for the U.S. Geological Survey, making major contributions to geologic and volcanic studies of the western United States and Hawaii.

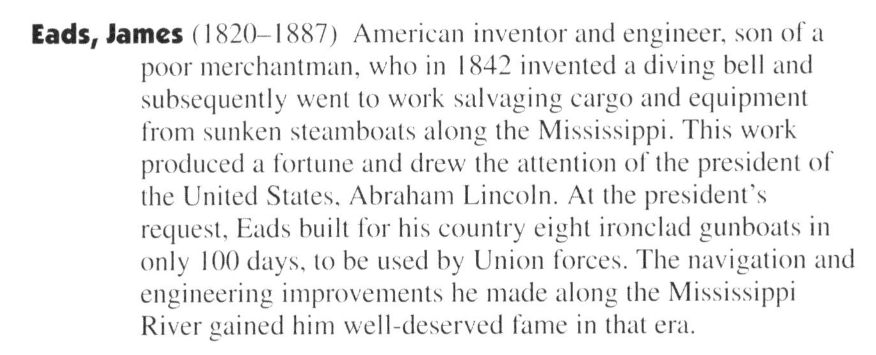

Eads, James (1820–1887) American inventor and engineer, son of a poor merchantman, who in 1842 invented a diving bell and subsequently went to work salvaging cargo and equipment from sunken steamboats along the Mississippi. This work produced a fortune and drew the attention of the president of the United States, Abraham Lincoln. At the president's request, Eads built for his country eight ironclad gunboats in only 100 days, to be used by Union forces. The navigation and engineering improvements he made along the Mississippi River gained him well-deserved fame in that era.

Sylvia Earle (Courtesy of NOAA/NURP)

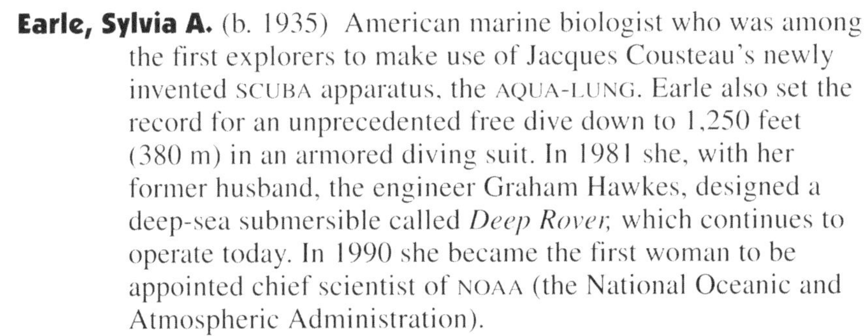

Earle, Sylvia A. (b. 1935) American marine biologist who was among the first explorers to make use of Jacques Cousteau's newly invented SCUBA apparatus, the AQUA-LUNG. Earle also set the record for an unprecedented free dive down to 1,250 feet (380 m) in an armored diving suit. In 1981 she, with her former husband, the engineer Graham Hawkes, designed a deep-sea submersible called *Deep Rover,* which continues to operate today. In 1990 she became the first woman to be appointed chief scientist of NOAA (the National Oceanic and Atmospheric Administration).

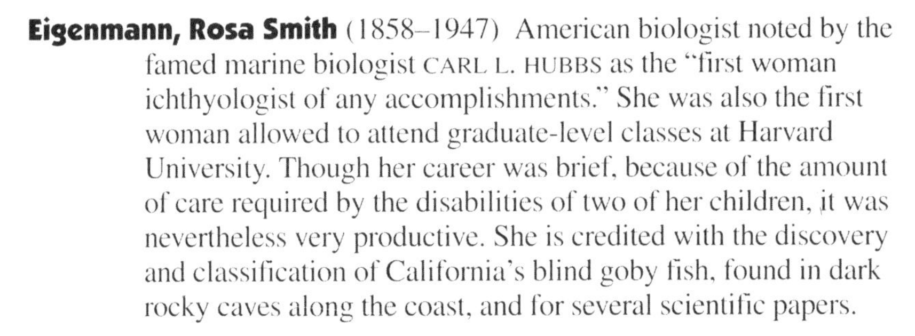

Eigenmann, Rosa Smith (1858–1947) American biologist noted by the famed marine biologist CARL L. HUBBS as the "first woman ichthyologist of any accomplishments." She was also the first woman allowed to attend graduate-level classes at Harvard University. Though her career was brief, because of the amount of care required by the disabilities of two of her children, it was nevertheless very productive. She is credited with the discovery and classification of California's blind goby fish, found in dark rocky caves along the coast, and for several scientific papers.

Ekman, Vagn Walfrid (1874–1954) Swedish oceanographer who helped found modern oceanography and invented many instruments to study ocean currents. He posed the question as to why the direction taken by drift ice seemed to differ from wind direction. His studies produced the familiar oceanographic terms EKMAN TRANSPORT, EKMAN LAYER, and EKMAN SPIRAL.

See also DEAD WATER PHENOMENON.

Eratosthenes (c. 276–c. 194 B.C.E.) Greek geographer, mathematician, and astronomer remembered for making the first calculations of the Earth's circumference, which he stated correctly to within 50 miles (80 km). He is also responsible for the first world map to show lines of latitude and longitude.

Ericson, Leif (Leiv Eiriksson) (c. 980–c. 1025) Norse explorer thought to be the first European, with his party, to set foot on the mainland of North America around 1000 C.E. He was the second son of ERIC THE RED. There are two scenarios as to what actually took place during the voyage that carried him to North America. One is that he while en route to Greenland he was blown off course and quite by accident found the shores of North America. The other is that after hearing accounts from an Icelandic explorer named Bjarni Herjulfsson that a land existed far to the west, Ericson bought Herjulfsson's ship, outfitted a party, and set off quite deliberately, and successfully, in search of it.

Ericsson, John (1803–1889) Swedish inventor who was the first to design propeller-driven ships. He moved to the United States in 1839 and designed novel underwater screw-driven canal ships, merchant steamers, and warships. In 1861–62 he designed and built the warship USS *Monitor*, which employed the screw for propulsion. Ericsson was a top-notch engineer who had a natural talent for the field. He was active in improving the application of compressed air for power, in hot-air engines, and in the railway locomotive.

Eric the Red (c. 950–c. 1001) Born Eric Thorvaldson, a redheaded Viking explorer credited with the colonization of Greenland. In about 985 he led a contingent of 25 ships, filled with colonists, toward the shores of Greenland. Only 14 made it. Two settlements, totaling around 450 people, were established. Documentation about the life of Eric the Red is found mainly in two writings from the 1100s or 1200s: *The Greenlanders' Saga* and *Eric's Saga.*

Ewing, William Maurice (1906–1974) American marine geologist known for his contributions to the knowledge of the ocean floor and the role he played in the development of the BATHYTHERMOGRAPH. His studies determined that the Earth's

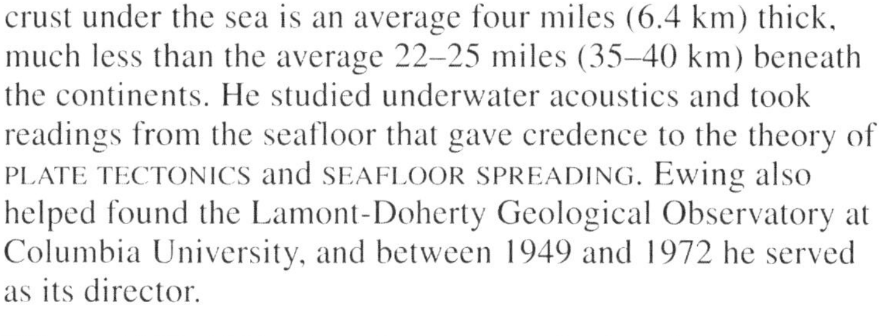

crust under the sea is an average four miles (6.4 km) thick, much less than the average 22–25 miles (35–40 km) beneath the continents. He studied underwater acoustics and took readings from the seafloor that gave credence to the theory of PLATE TECTONICS and SEAFLOOR SPREADING. Ewing also helped found the Lamont-Doherty Geological Observatory at Columbia University, and between 1949 and 1972 he served as its director.

Ferrel, William (Courtesy of NOAA)

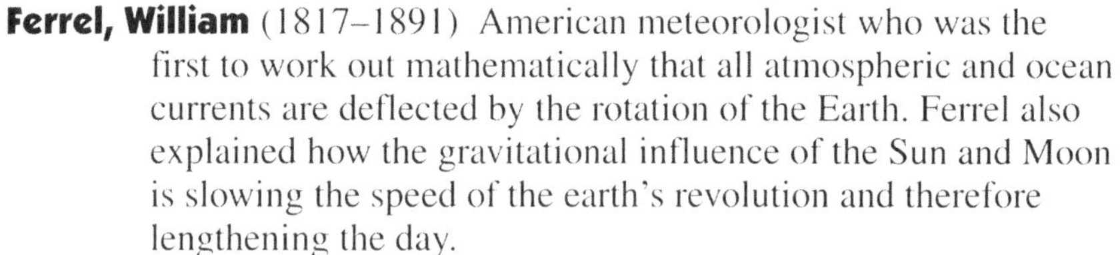

Ferrel, William (1817–1891) American meteorologist who was the first to work out mathematically that all atmospheric and ocean currents are deflected by the rotation of the Earth. Ferrel also explained how the gravitational influence of the Sun and Moon is slowing the speed of the earth's revolution and therefore lengthening the day.

Findlay, Alexander George (1812–1875) English geographer and engraver who, in 1843, published the *Modern Atlas*, engraved by Findlay himself. He was the son of Alexander Findlay, also an engraver. In the 1850s Findlay took over the operation of Laurie and Whittle, a prominent map publishing business of England.

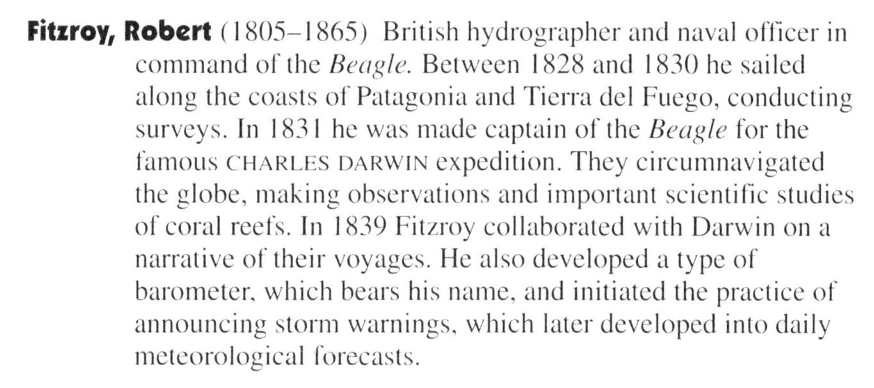

Fitzroy, Robert (1805–1865) British hydrographer and naval officer in command of the *Beagle*. Between 1828 and 1830 he sailed along the coasts of Patagonia and Tierra del Fuego, conducting surveys. In 1831 he was made captain of the *Beagle* for the famous CHARLES DARWIN expedition. They circumnavigated the globe, making observations and important scientific studies of coral reefs. In 1839 Fitzroy collaborated with Darwin on a narrative of their voyages. He also developed a type of barometer, which bears his name, and initiated the practice of announcing storm warnings, which later developed into daily meteorological forecasts.

Flamsteed, John (1646–1719) English astronomer who, in 1675, became the first director of the Royal Greenwich Observatory in England, which was founded the same year. Throughout his entire life Flamsteed suffered from sickliness and frailty, yet, acting as the first royal astronomer, he became obsessed with determining accurate stellar and lunar positions as an aid to navigation. Flamsteed's unique development was to employ a telescopic sight on his seven-foot (two-meter) sextant to make observations.

Flinders, Matthew (1774–1814) British explorer, hydrographer, and self-taught navigator who, from 1801 to 1803, circumnavigated Australia, making accurate chartings of the coast. During the trip home to England, his ship became wrecked on the then largely uncharted Great Barrier Reef. His 700-mile (1,127-km) return to Australia in the ship's cutter was quite a notable feat in itself. In 1810, after being released from imprisonment at French-occupied Mauritius, when Britain and France were at war, Flinders returned to England and wrote *A Voyage to Terra Australis* (1814); thus he finally received due credit for the charting of Australia. The Flinders River, Queensland, and the Flinders Range, South Australia, are named in his honor.

See also BASS, GEORGE.

Forbes, Sir Edward (1815–1854) British naturalist who did much to promote interest in marine biology. He was the first to study the distribution of life throughout different depths of the ocean, observing that at depths beyond which light could penetrate no plant life existed. He studied the organisms of the Aegean Sea and encouraged biological explorations of the deep oceans; however, he held the incorrect assumption that life could not exist in the darkness of the depths at around 1,600 feet (488 m).

See also ROSS, SIR JOHN.

Forchhammer, Johann Georg (1794–1865) Danish geologist, chemist, and director of the Science Museum in Copenhagen, Denmark, who identified the individual elements that seawater comprises. In 1865, he determined that there is a constant ratio of salts found in seawater taken from different locations. This assertion of a constant ratio is known as Forchhammer's principle.

Franklin, Benjamin (1706–1790) American printer, author, diplomat, philosopher, and scientist who published in 1769 the first ocean charts of the Gulf Stream to aid the passage of ships across the Atlantic.

Franklin, Sir John (1786–1847) British explorer, naval officer, and pioneer of English exploration of the Arctic who died while seeking the NORTHWEST PASSAGE. In 1803 he served as midshipman on the voyage MATTHEW FLINDERS (his uncle) made around Australia. In 1819 he led his first Arctic expedition to the mouth of the Coppermine River, and he returned to the Arctic on a second voyage in 1825–26. In 1845, commanding the *Erebus*, Franklin led the most well equipped expedition ever to enter the Arctic. This time he sought the Northwest Passage. His party never returned. Distraught by his absence, his wife launched a search effort led by SIR ROBERT MCCLURE, but Franklin was never found; only signs of his passing were. Evidence from McClure's search enabled Franklin's route to be reconstructed.

Fresnel, Augustin-Jean (1788–1827) French engineer and physicist credited for his work in double refraction, which resulted in the creation of the Fresnel lens, used most famously in the operation of lighthouses. The lens is able to concentrate light and focus a beam to shine for great distances.

Frobisher, Sir Martin (c. 1535–1594) English explorer known for making three voyages west in search of the NORTHWEST PASSAGE from Europe to Asia. Frobisher's first voyage, in 1576, was the first English attempt to find a passage to the Orient. On his return from his next two, in 1577 and again in 1578, he took home raw ore thought to be ribboned with gold. Ultimately, all was found to be worthless. Though his search for gold and the Northwest Passage was a failure, his skills at sea earned him a reputation as one of the greatest navigators of his time. Frobisher Bay of Baffin Island, Canada, honors his name.

the *Clermont*

Robert Fulton steamboat, the *Clermont*

Fulton, Robert (1765–1815) American engineer, artist, and inventor responsible for designing the *Clermont* (1807), the first commercially successful steamboat, though not the first steamboat built. Fulton received a patent for its construction. About 1797, he made important designs for the development of the submarine, and during an extended tour in France he developed the *Nautilus,* an underwater diving boat; however, by 1806 lack of funding sent him home to the United States. With the help of the American engineer NICHOLAS J. ROOSEVELT, he also designed the *New Orleans,* the first steamboat on the Ohio and Mississippi Rivers, and the warship *Fulton the First,* which defended New York Harbor. He died before the *Fulton* was completed.

Gama, Vasco da (c. 1469–1524) Portuguese navigator who led the first European expedition to India by sailing around the Cape of Good Hope, Africa, in 1497. However, the Indian ruler was not impressed with da Gama's gifts, and the captain returned to Lisbon in 1499. In 1502 he again sailed to India, this time in the name of revenge. Da Gama's party slaughtered countless innocent Muslims and Indians, avenging the murders of Portuguese explorers left there by Cabral in 1500. In 1503 he retired from the sea. Then in 1524 he was named viceroy of India. Da Gama set sail one last time and died in India.

Gellibrand, Henry (1597–1636) English astronomer and navigator best known for his discovery of the change in magnetic declination that occurs over time. Based on celestial observations recorded in his *Appendix Concerning Longitude* (1633), Gellibrand's discovery laid to rest the hope that a single consistent declination could be used to determine longitude. In 1627 Gellibrand became professor of astronomy at Gresham College, London.

Gerard, John (1545–1612) British botanist worthy of mention for his work *The Herball,* or *General Histoire of plantes* (1597). It was the most comprehensive botanical volume of his day, identifying more than 1,000 species of plants.

Gesner, Conrad (1516–1565) Natural historian noteworthy for his work *Historia animalium* (1551–58), in which he attempted to present alphabetically all information that was known about animals up to that time. He also added more than 500 species of plants not addressed by early naturalists.

Gilbert, Sir Humphrey (c. 1539–1583) English navigator and half brother of Sir Walter Raleigh, who led movements to seek the NORTHWEST PASSAGE and to colonize the Americas. He was responsible for the annexation of Newfoundland by England in 1583.

Gill, Theodore Nicholas (1837–1914) American ichthyologist and professor of zoology at George Washington University,

Washington, D.C. He was called by some the "master of of taxonomy" and is probably best known for his detailed classification of fishes. His writings proved priceless contributions to the field of ichthyology.

Goode, George Brown (1851–1892) American ichthyologist, natural scientist, and museum administrator who was among the foremost museologists (museum professionals) of his time. He was educated at Wesleyan University, Connecticut, and at one point studied under LOUIS AGASSIZ at Harvard University. Beginning in 1872, Goode worked for the United States Fish Commission in Eastport, Maine, doing summer fieldwork for them, the Smithsonian, and Wesleyan. In 1877 he joined the Smithsonian full-time; he eventually became curator. He was also curator of the newly formed U.S. Natural History Museum. Goode was a devoted advocate of natural science, especially that which involved fishes. He knew that a broader view of science could only improve the lives of the human race. Among his extensive writings are *American Fishes* (1888) and the monumental *Oceanic Ichthyology* (1895).

Gray, Robert (1755–1806) American sea captain and explorer famous as the first American to circumnavigate the globe. In 1787, commanding the *Lady Washington,* one of a two-vessel team, he sailed from Boston on a fur trading expedition toward the North Pacific. He was later put in charge of the voyage and given command of the second ship, the *Columbia.* In 1790 the party returned by rounding the Cape of Good Hope. He again set sail in 1792, along the same route, and came upon the mouth of the Columbia River, named after his ship. Gray's discovery of the river laid the basis for future claim to the Oregon Territory.

Haeckel, Ernst (1834–1919) German naturalist and professor of zoology at Jena, Germany, best known as the first to map a genealogical tree of animals and as a strong supporter of CHARLES DARWIN's theory of natural selection. He coined the words *phylum, phylogeny,* and *ecology.* Haeckel worked on invertebrates such as poriferans (sponges) and annelids (worms) and identified nearly 150 new species of radiolarians while conducting research in the Mediterranean.

Hakluyt, Richard (c. 1552–1616) English geographer who wrote extensively on navigation and exploration. His early claim to fame was his first book, *Divers Voyages Touching the Discovery of America and the Islands Adjacent* (1582). In it he details English exploration of the New World. Later he published his better-known work, *Principal Navigations, Voyages, and Discoveries of the English Nation* (1589). He revised and expanded it to three volumes, in 1598–1600, weaving together fact and fiction about sea captains, their voyages to unexplored regions, and their sightings of sea monsters. In 1846, in England, the Hakluyt Society, which publishes books on exploration, discovery, and travel, was founded in his honor.

See also Useful websites in the Appendixes.

Haldane, John Scott (1860–1936) English physiologist best known for his study of respiratory function. His initial goal was to learn more about the mining and industrial diseases caused by poor ventilation. Haldane's research led to his working out a method for slow decompression on ascent from deep free-dives. His work proved essential to the design of ventilation systems aboard submarines and of diving equipment.

Halley, Sir Edmund (1656–1742) British astronomer most famous for the comet that bears his name, Halley's comet. However, among his many scientific achievements, he made a voyage that is believed to have been the first truly science-oriented expedition. In 1698 he set sail into the Atlantic to study the variations of a magnetic compass and traversed as far south as 52° S. He also voyaged to Saint Helena and recorded significant findings that added to the knowledge of the behavior of TRADE WINDS. He wrote *The Three Voyages of Edmund Halley in the Paramore, 1698–1701.*

Happel, Eberhard Werner (1647–1690) German writer known for his work *Groste Denkwurdigkeiten der Welt* (World's greatest curiosities [1683–91]), in which he proposed an explanation of tides and included the second known chart illustrating the oceanic circulation of the globe. The book touched on many subjects and grew so popular that fakes and sequels emerged during the decades that followed.

Harrison, John (1693–1776) English horologist and inventor who, in 1735, as the result of an incentive offered to the public 19 years earlier by the British Board of Longitude and amid the pressure of competition, invented the first accurate marine chronometer. His fourth design was put to the test during a five-month sea trial in 1761–62 and was found to have an error of only 1.25 minutes, surpassing the board's requirements and earning him the £20,000 prize, which incidentally he took 10 years to collect.

Hays, James Douglas (b. 1933) American geologist who in 1971 published a report that eight species of radiolarian had thrived and then become extinct within the last 2.5 million years. This finding was based on 28 deep-sea piston corings of bottom sediment made at various depths. He deduced that a change in the Earth's magnetic field brought about their demise.

Heezen, Bruce Charles (1924–1977) American oceanographer at Lamont-Doherty Earth Observatory. While in collaboration with Marie Tharp, he is best known for charting the ocean floors and making new geological discoveries in the process, for example, of the deep rift that traverses the spine of the Mid-Atlantic Ridge. In 1952, using records from a 1929 Grand Banks earthquake that caused numerous failures in communication cables, Heezen discovered oceanic turbidity currents. He was the first to make known their existence and to explain the importance of their role within the ocean.

Henry the Navigator (1394–1460) Portuguese prince, third son of King John I and Queen Philippa of Portugal, who was a great promoter of scientific navigation and exploration. It was his influence that caused Portugal to rise as a leader in exploration among all the European nations. Prince Henry established a school of navigation and an observatory at Sagres, Algarve, Portugal, and he sponsored many exploratory expeditions. Cartographers, astronomers, and mathematicians of variety of different nationalities flocked to his aid. The West African coast was explored in detail, and the way was prepared for the discovery of the sea route to India. After his death, voyage after voyage was made because of the navigational knowledge gained under his direction, including the historic expeditions led by VASCO DA GAMA and BARTOLOMEU DIAS.

Hensen, Victor (1835–1910) German scientist who broke new ground for techniques used in plankton study, such as the implementation of a sampling net of his own design. While aboard the ship *National,* he was the first to conduct large-scale plankton study properly; Hensen concluded that plankton is as abundant in Earth's deep sea polar regions as in its tropical zones.

Henson, Matthew Alexander (1867–1955) African American explorer who, along with ROBERT E. PEARY, was the first to reach the North Pole in 1909. Henson went to sea at a young age, 13, and at 21 he met Peary. For the next 22 years Henson served as Peary's personal assistant. He stuck by Peary's side through each expedition, his knowledge and experience equaling Peary's. It is quite possible that, if not for Henson, Peary might not have accomplished the historical 1909 expedition to the pole. Henson was a dog musher (sledger) of unrivaled skill, and much of the time he was the one breaking trail, leading the way northward. Sadly, because of the racial discrimination in America at that time, not until his later years did Henson receive the honors he deserved for his role in the expeditions. He wrote about his adventure to the North Pole in the book *A Negro Explorer at the North Pole* (1912), which can be found reprinted at bookstores today. Bradley Robinson wrote a biography of Henson, *Dark Companion,* which was published in 1947.

Hess, Harry Hammond (1906–1969) American geologist who discovered that the rock of the ocean floors is younger than the rock of the continents, a finding that led to the theory of the phenomenon of SEAFLOOR SPREADING. His findings contributed greatly to the theories of continental drift and plate tectonics. Also, while working for the U.S. Navy during World War II, Hess discovered the existence of flat-topped seamounts, which he labeled GUYOTS.

Heyerdahl, Thor (1914–2002) Norwegian anthropologist and explorer who is best known for his explanation of the possibility that early humans may have navigated as nomads over vast expanses of water. In 1947 Heyerdahl and five others set sail from Peru on a balsa raft, the *Kon Tiki*, modeled from ancient Peruvian rafts. They traveled 4,300 miles (6,920 km) across the South Pacific and landed successfully at the Tuamotu Islands in an attempt to prove his theory that Polynesians could have migrated from South America. He made two other successful voyages on a primitive Peruvian raft, one to the Galápagos in 1954 and another to Easter Island and the East Pacific regions in 1955–56, putting to the test the same theory—that the ancients could have possessed great skills of seamanship and thus populated the world. He also tested a theory that early Egyptians may have founded the Aztec and Inca cultures. Using a reed boat, he sought to travel across the Atlantic from North Africa. The first attempt sank, but the second landed him safely at Barbados after only 57 days at sea. Heyerdahl's expeditions won him fame and several honors. He wrote *Kon-Tiki: Across the Pacific by Raft* (1947), *Ra Expeditions* (1971), and *Early Man and the Ocean* (1979).

Hipparchus (c. 180–125 B.C.E.) Greek astronomer who discovered the PRECESSION OF THE EQUINOXES, estimated the Earth's distance from the Sun and the Moon, and used mathematics to locate geographical points by latitude and longitude.

Hjort, Johan (1869–1948) Norwegian biologist whose studies of the Mediterranean and the North Atlantic formed the basis for the classic book *The Depths of the Ocean* (1912), in which he wrote, among other topics, about mesopelagic and bathypelagic organisms.

Holland, John Philip (1841–1914) Irish-American inventor who designed and built the first practical submarine and is therefore responsible for its development. He arrived in the United States in 1873, plans in hand, but it was not until 1888 that the U.S. Navy finally agreed to give them a look. Holland built the navy a boat; however, it failed because of all the changes they made to his original plans. Perturbed, the inventor built his own boat, which he christened the *Holland.* It was a success, of course, and the navy purchased the *Holland* in 1900. Holland's firm, the Electric Boat Company, continued to build many subs for the navy and is still in operation today as the Electric Boat Division of General Dynamics Corporation.

Hooke, Robert (1635–1703) English experimental scientist whose interests knew no bounds. He delved into astronomy, physics, chemistry, biology, geology, and even architecture. He is probably best known for his development of the equation of elasticity called Hooke's law. He collaborated with many prominent scientists such as ROBERT BOYLE and ANTONI VAN LEEUWENHOEK and assisted them in their work. Hooke designed such inventions as a clock spring that improved the accuracy of timekeeping; he built the Gregorian reflecting telescope and a microscope with which he discovered new plant cells; and he built a barometer, an anemometer, and a glass ball depth-sounding device, which he could never quite get to work properly. Though Hooke made valuable contributions to nearly every field of science, many of his experiments lie unfinished, for he sometimes lacked the diligence to pursue his every inspiration.

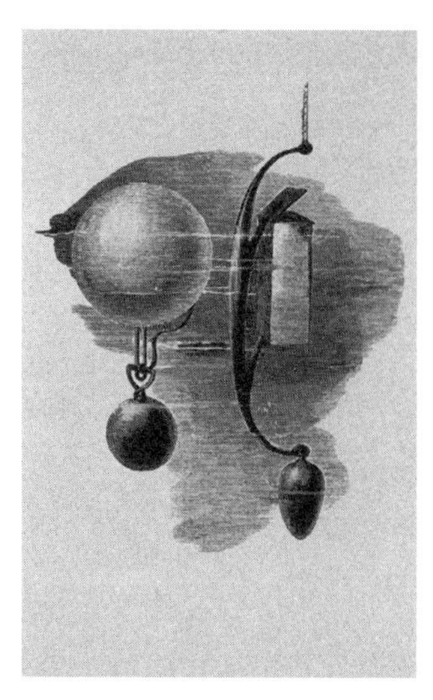

Robert Hooke glass ball sounding machine
(Courtesy of NOAA)

Hubbs, Carl L. (1894–1979) Famous American marine biologist of many accomplishments. From 1920 to 1944 he acted as curator of fishes for the University of Michigan Museum of Zoology, and from 1928 to 1930 he served as secretary of the American Society of Ichthyologists and Herpetologists. He was supervisor of fisheries for the Michigan Department of Natural Resources IFR (Institute of Fisheries Research) from 1930 to 1935; in 1944 he was appointed professor of biology at the Scripps Institution of Oceanography. In 1947 Hubbs helped establish the California Cooperative Oceanic Fisheries Investigation, and between 1962 and 1971 he served as vice president of the San Diego Zoological Society. Hubbs, a leader in his field, wrote extensively on biological science and led many research trips all across North and South America accompanied by his wife, Laura Hubbs. In 1954 he discovered a colony of Guadalupe fur seals, a species thought to be extinct.

Hudson, Henry (d. 1611) English navigator who, between 1607 and 1611, made four voyages of discovery to seek out a NORTHWEST PASSAGE. Using only a single ship and crew for each trip, Hudson never established the much-desired northern trade route between Europe and Asia, but he did sail farther north than any other who had gone before him. He discovered three waterways, which were named for him: the Hudson Bay, Hudson River, and Hudson Strait. In 1611, some of his crew mutinied because of the conditions of hardship dictated by the

isothermal line

cooler warmer

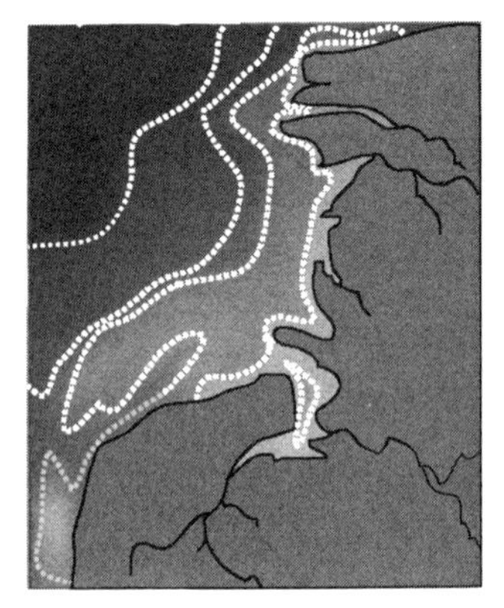

isothermal lines devised by Friedrich von Humboldt

Arctic climate. Hudson, his son, and a few other crew members were set adrift in a lifeboat. The mutineers made it safely back to England but were promptly imprisoned. Hudson and his castaways were never seen again.

Humboldt, Friedrich von (1769–1859) German explorer, scientist, climatologist, and geographer who helped found methods of modern geography. Humboldt was a pioneer in correlating climate to the geographical locations of different flora and was the first to draft a map showing lines that connect geographic points of similar temperature, called isothermal lines. He also worked on geomagnetism, which prompted him to the fine-tuning of geographic measurement. He led many adventurous explorations throughout Venezuela, collecting data on oceanographic, zoological, and geographic phenomena.

Huxley, Sir Julian Sorell (1887–1975) English biologist and writer who helped encourage the investigation of biological science among the public, writing on scientific issues in terms people could understand and grow excited about. Huxley, the grandson of THOMAS H. HUXLEY, became professor of zoology at King's College, London (1925–27) and at the Royal Institution (1926–29) and was secretary to the Zoological Society of London (1935–42). He was the first director-general of UNESCO (the United Nations Educational, Scientific and Cultural Organization) (1946–48). Included in his writings are *Essays of a Biologist* (1923) and *Evolution: The Modern Synthesis* (1942).

Huxley, Thomas Henry (1825–1895) English zoologist, writer, and defender of Darwin's theory of evolution. Signing on as surgeon, Huxley voyaged into the southern seas of Australia and New Guinea aboard the HMS *Rattlesnake,* a frigate of the Royal Navy. When not attending to the responsibilities of ship's physician, he studied marine invertebrates, mostly cephalopod mollusks, tunicates, and coelenterates, a practice that earned for him a position within the prestigious scientific community of England. Among his outstanding writings are *Man's Place in Nature* (1863) and *A Manual of the Anatomy of Invertebrated Animals* (1877).

d'Iberville, Pierre Le Moyne, sieur (1661–1706) French-Canadian explorer and naval officer who founded the French province of Louisiana. He led sea and land missions against the British-run territories of the Hudson Bay area in an attempt to oust them from the land and eventually succeeded. In 1699 he became the first European to enter the mouth of the Mississippi by way of the Gulf of Mexico, and he explored its banks as far north as modern Cairo, Illinois. He sighted Louisiana's Lake Pontchartrain, which he named after Count Pontchartrain, head of the French navy at that time.

Iselin, Columbus O'Donnell (1904–1971) American oceanographer and developer of a type of bathythermograph who carried out important research concerning ocean salinity and temperature, and acoustics and the oceanographic characteristics of the GULF STREAM and western North Atlantic Ocean. Between 1932 and 1970 he served at Woods Hole Oceanographic Institution as captain of the research vessel *Atlantis* and as director after 1940. He also taught at the Massachusetts Institute of Technology between 1959 and 1970 and at Harvard University between 1960 and 1970. In 1954 he was elected to the National Academy of Sciences.

Jeffreys, Sir Harold (1891–1989) British geophysicist, mathematician, and astronomer who is known for his forceful opposition to ALFRED WEGENER's theory of CONTINENTAL DRIFT. It was because of Jeffreys and his influential work *The Earth: Its Origin, History and Physical Constitution* (1924) that Wegener's correct theory lay on the back porch for as long as it did.

Jeffreys, John Gwyn (1809–1885) Welsh lawyer and conchologist who, along with WILLIAM B. CARPENTER and CHARLES W. THOMSON, made dredging expeditions aboard the ships *Shearwater, Porcupine,* and *Lighting* in the eastern North Atlantic and Mediterranean Sea. They recorded temperature observations that provided proof of the existence of a circulation system below the surface and collected new organisms from as deep as 12,000 feet (3,657 m). These surveys caused the Admiralty and the Royal Society to mount the famous *Challenger* expedition. Jeffreys is the author of the five-volume *British Conchology* (1862–69).

Joly, John (1857–1933) Irish geologist who in 1888 tried to calculate the age of the Earth by measuring the salinity of the oceans. That method, however, proved inaccurate, and he turned to using radioactive decay in the Earth's crust instead. With this method, in 1913 he estimated that the beginning of the Devonian period occurred no less than 400 million years ago, a date that agrees closely with modern estimates. Today's calculations put the Earth's age at about 4.6 billion years.

Jussieu, Antoine Laurent de (1748–1836) French plant taxonomist best remembered for his introduction of a botanical classification that organized plants by using a more detailed view of their characteristics, which differed from the Linnean system, which used only the most basic traits. Jussieu's most famous work is *Genera plantarum* (1789). Three-quarters of his 100 plant family classifications are still used today.

Kircher, Athanasius (1602–1680) Jesuit priest who divided his devotion between God and science. In his great two-volume work *Mundus subterraneus* (1678), Kircher produced one of the earliest charts of global ocean circulation. On it he demonstrated broad western flow in the Indian and Pacific Oceans and greater detail in the better-known Atlantic Ocean, where he showed gyres and whirlpools. He also presented a curious wild speculation that water flowed into the center of the Earth at the North Pole and was released rhythmically at the South Pole, explaining the regularity of the rise and fall of the tides.

Knudsen, Martin Hans Christian (1871–1949) Danish physicist and oceanographer who developed an effective way to measure salinity in a liquid; that method, together with improved sampling equipment, made the detailed study of deep-water circulation finally possible. He also established the Knudsen flow, the flow of very-low-density gas inside a container.

Kotzebue, Otto von (1787–1846) German naval officer and explorer in the service of Russia who participated in ADAM JOHANN VON KRUSENSTERN's first Russian circumnavigation of the globe and then commanded two circumnavigations of his own, the first in 1815–18 and the second in 1823–26. During his explorations many deep-sea temperature observations and scientific data were recorded. While trying to establish a passage through the Arctic Ocean, Kotzebue discovered a sound off the northern coast of Alaska, which is named for him. Kotzebue wrote the three-volume *A Voyage of Discovery* (1821) and the two-volume *A New Voyage around the World* (1830, reprinted in 1967).

Krümmel, Otto (1854–1912) German geographer and oceanographer who took part in many research voyages and contributed greatly to knowledge of oceanic circulation, such as detailing characteristics of the FALKLANDS CURRENT and the North Equatorial Countercurrent. Krümmel's celebrated work *Handbuch der Ozeanographie* (1887) stood as the standard reference on physical oceanography of the day.

Krusenstern, Adam Johann, baron von (1770–1846) Russian admiral famous for commanding the first Russian voyage around the world, in 1803–06, which significantly expanded knowledge of the North Pacific. He made an account of his expedition in his *Voyage round the World* (1809–13).

Kuenen, Philip Henry (1902–1976) Dutch geologist who helped found modern sedimentology. He studied coral atolls and oceanic plains and showed how ocean currents were responsible for the formation of deep clefts in continental slopes. Kuenen's most popular work was his *Marine Geology* (1950).

Lamarck, Jean-Baptiste-Pierre-Antoine de Monet, chevalier de (1744–1829) French naturalist best known for his study of invertebrates and his theory of evolution called lamarckism, which held that life-forms evolved through hereditary traits that changed gradually over the eons as they adapted to the shifting environment. He coined the terms *biology, Invertebrata, Infusoria, Annelida, Crustacea, Arachnida,* and *Tunicata* and is deemed the creator of invertebrate paleontology. His theories drew harsh criticism from certain colleagues of well-established repute, which hobbled his efforts at prominence and acceptance, but naturalists such as CHARLES DARWIN acknowledged him as a great zoologist. Lamarck's writings include the three-volume *Flore françoise* (1778), two-volume *Philosophie zoologique* (1809), and seven-volume *Histoire naturelle des animaux sans vertèbres* (1815–22).

Lapérouse, Jean-François de Galaup de (1741–1788) French navigator who went in search of the NORTHWEST PASSAGE and never returned. His path can be traced from his departure from France in 1785. He sailed around Cape Horn, the tip of South America, and on to the Hawaiian Islands, Alaska, and the Philippines. In 1787 he discovered La Pérouse Strait and landed at Kamchatka, where he sent a party overland to France with his journals in their care. He then sailed to Australia and departed Botany Bay in 1788, after which his trail suddenly grew cold. In 1826 parts of what is believed to be his ship were found east of Australia on an island of Vanuatu (formerly New Hebrides).

Laplace, Pierre-Simon Laplace, marquis de (1749–1827) French astronomer and mathematician best known for his theory on stable planetary orbits of our solar system and for his literary work *Mécanique céleste* (Celestial mechanics, 1799–1825), which systematized all his mathematical work done on gravity. Also, it is said that de Laplace is responsible for launching scientific study of the deep sea by using a mathematical formula to calculate the depth of the Atlantic Ocean. He gathered records of tidal swings on the Brazilian and African coasts and then applied his formula, putting the depth at around 13,000 feet (3,962 m), a figure that later soundings proved quite accurate.

Leeuwenhoek, Antoni van (Antony) (1632–1723) Dutch microscopist (microscope professional) known for crafting a magnifying device that far exceeded all others of his time. Van Leeuwenhoek, a tradesman, was a self-taught scientist with an insatiable craving for knowledge. He used great care and skill to grind small double-convex lenses that he mounted on a brass plate, which was held close to the eye in good lighting. Separate screws would adjust focus. Van Leeuwenhoek's lenses could magnify his subject more than 200 times, whereas his colleagues' lenses could enlarge only 20 to 30 times. He observed mineral crystals, fossils, and plant and animal tissue. He discovered bacteria and protozoa and stated that mussels were produced from tiny eggs, not by spontaneous generation (the spontaneous origination of organisms directly from inorganic materials). He was the first to see FORAMINIFERA. Van Leeuwenhoek kept the method of his lens crafting secret, however, and the next detailed observation of bacteria was not made until the 19th century, when the common compound microscope was finally improved.

Lenz, Heinrich Friedrich Emil von (1804–1865) Russian physicist and professor at the University of Saint Petersburg who described the circulation of ocean currents and the upwelling of cold water from the Poles toward the equator. He also did work in magnetism and formulated Lenz's law, which shows that an electrical current induced by a changing magnetic field creates its own magnetic field which opposes the magnetic field that created it.

Leonardo da Vinci (1452–1519) Italian engineer, scientist, artist, architect, and musician, who was famous for the extent of his scientific analysis. Many of his notes dealt with oceanographical subjects. It is known that, at the request of the Venetian senate, he was the designer of the first snorkel, which consisted of a hollow tube connected to a leather helmet.

Linnaeus, Carolus (Carl von Linné) (1707–1778) Swedish botanist known as the founder of the modern botanical taxonomic classification system, which followed an early classification system created by JOHN RAY. He traveled extensively, enriching his knowledge of plants and, of lesser interest, animals and minerals. His most famous works are his *System naturae* (1735), *General planterium* (1737), and *Species plantarum* (1753). In 1742 he became professor of botany at University of Uppsala in Uppsala, Sweden, north of Stockholm.

Linschoten, Jan Huygen van (1563–1611) Dutch explorer, merchant sailor, and author who participated in two voyages in search of NORTHEAST PASSAGE between Western Europe and China. In 1594 he accompanied WILLEM BARENTS on Barents's second voyage to find the passage, but they made it only as far as the Kara Sea. In 1595 van Linschoten set forth on yet another attempt to find the passage for Holland, but the fleet of seven ships was turned back by the impassable ice pack. He wrote about his trip with Barents in his *Journalen* (1601). Van

Linschoten also wrote *Itinerario* (1595–96), in which he described his famous trip to India (1583–88) and the lands bordering the western Pacific and Indian Oceans. *Itinerario* included two plates that were the first published renderings of Saint Helena.

Lubbock, Sir John William (1803–1865) English astronomer, mathematician, and banker known for his study of the tides and lunar theory. He also developed calculations for the orbital paths of planets and comets. Certain aspects of modern knowledge of these subjects can be attributed to Lubbock's work. A small lunar crater is named for him.

Magellan, Ferdinand (c. 1480–1521) Portuguese sea captain who led the first expedition to sail around the world in order to prove the Earth is not flat. Magellan's original goal was to find a shorter route to the Spice Islands. After years of preparation and under the suspicious eyes of the Spaniards, who were financing his mission, Magellan set sail from the south of Spain in September 1519. He commanded a total of 241 men and a fleet of five ships, the *Concepción, San Antonio, Santiago, Trinidad,* and *Victoria.* In 1520 he was the first to lead an expedition through what he named the Straits of Magellan off the southern tip of South America, a trip during which he named the ocean *Pacific,* meaning "peaceful." Magellan did not live to finish the voyage; he died in a battle with Philippine Islanders in April 1521. From the start, the expedition had encountered much hardship, and in the end only the *Victoria,* captained by JUAN SEBASTIAN DEL CANO, succeeded in rounding the globe and returning to Spain in 1522 with only 17 surviving crewmen. Facts recorded in the journal of one of the surviving crewmen, Antonio Pigafetta, stated that the voyage covered 14,460 leagues (50,610 miles [81,449 km]) and indicated that Magellan's courage and navigational skills had produce their success. However, it was the returning del Cano to whom the credit for the voyage was given. Nonetheless, it was Magellan's hard work and perseverance that made his dream of circumnavigating the globe a reality. And although he did not succeed in finding a shorter route to the Spice Islands, the data his voyages provided contributed greatly to the knowledge about the Earth's oceans.

Makarov, Admiral Stepan (1849–1904) Russian admiral and scientist who became well known for his oceanographic work aboard the research vessel *Vitaz* and for his work in founding the International Council for the Exploration of the Sea.

Marsigli Ferdinando, count de Luigi (1658–1730) Italian naturalist and geographer who divided his occupation between those of soldier and scientist. He recorded measurements of pressures and temperatures at various depths in the oceans, measured the speed and size of rivers, made astronomical observations, and collected fossil specimens. In 1715 Marsigli Ferdinando founded the Institute of Sciences, Bologna. His major works include *Osservazioni interne al Bosforo Tracio* (1681) and *Histoire physique de la mer* (1725).

Maury, Matthew Fontaine (1806–1873) American naval officer and oceanographer who conducted systematic studies of the ocean and subsequently established the first shipping lanes of the sea. After a voyage around the world, Maury concluded that seafarers needed better knowledge of navigation and meteorology to increase the efficiency of shipping. He organized a world conference that took place in Brussels in 1853. Out of this meeting the International Hydrographic Bureau, now the IHO (International Hydrographic Organization), took shape, with the purpose of establishing international standards for easier and safer navigation at sea and promoting understanding and protection of the marine environment. Maury also created the first bathymetric chart of the North Atlantic Current. His classic *Physical Geography of the Sea* was published in 1855.

See also Useful websites in the Appendixes.

Mayr, Ernst Walter (b. 1904) German-born American zoologist who between 1928 and 1930 made three expeditions to the Solomon Islands and New Guinea to formulate a theory in keeping with CHARLES DARWIN's theory of evolution. He stated that new species arise when an existing species migrates from its natural habitat and then evolves in order to adapt to the new habitat. Mayr's works include *Animal Species and Evolution*

(1963), *Evolution and the Diversity of Life* (1976), and his monumental *Growth of Biological Thought* (1982).

McClure (M'Clure), Sir Robert John le Mesurier (1807–1873) British explorer who, between 1850 and 1854, led the first expedition that proved the existence of the NORTHWEST PASSAGE while in search of the ill-fated party of SIR JOHN FRANKLIN. Franklin had set forth to achieve that very objective five years earlier. McClure approached the Arctic from the west, sailed across the Beaufort Sea, and became stuck in the ice near the northern coast of Banks Island. Shortly before starvation took them all, a rescue party arrived in 1853. McClure then directed a land expedition to Viscount Melville Sound, a point reached by SIR WILLIAM E. PARRY, who had arrived there from the east in 1819–20.

Menard, Henry William (1920–1986) American marine geologist, theorist, entrepreneur, and politician best known for promoting the theory of PLATE TECTONICS before its acceptance into scientific circles. Between 1949 and 1978, Menard conducted extensive fieldwork involving 20 oceanographic expeditions and almost 1,000 scuba dives. In 1955 he joined the Scripps Institution of Oceanography as associate professor of geology, and in 1961 he became a professor at the University of California, San Diego. Menard was an adviser for the Office of Science and Technology in Washington, D.C., and a consultant on the laying of seafloor cable for AT&T (American Telephone & Telegraph). He was elected to the National Academy of Sciences in 1968. His works include *Marine Geology of the Pacific* (1964), *Anatomy of an Expedition* (1969), *Science: Growth and Change* (1971), and *Geology, Resources, and Society* (1974). After his death in 1986 two of his books, *Islands* and *Ocean of Truth: A Personal History of Global Tectonics,* were published in the same year.

Mercator, Gerardus (1512–1594) Flemish geographer, mapmaker, and mathematician who constructed in 1569 his famous map of the world, the Mercator Projection Map, designed specifically for use in navigation. The projection is valuable to navigators because on such a map straight lines between any two points show a constant compass direction for the course of a ship. On his map, the equatorial zones were depicted accurately but regions in the high latitudes were greatly distorted (which is what happens when for the purpose of global measurement a spherical Earth is projected on flat paper). The loyal rendering of directional lines was valued as priceless for navigation. Also, Mercator was the first to apply to a collection of maps or charts the term *atlas,* in Greek mythology the name of the Titan Atlas, who was condemned to hold the Earth on his shoulders.

Monts, Pierre de Gua, sieur de (c. 1560–c. 1630) French explorer who sailed for America in 1604 with Jean de Biencourt de Poutrincourt and SAMUEL DE CHAMPLAIN, with whom he explored the Bay of Fundy and settled at the mouth of the Saint Croix River. Sieur de Monts, whose given name was Pierre du Guast, was made a lieutenant general by King Henry IV of France and awarded the governorship of the fur trading post of Acadia, with the condition that he settle the area.

Müller, Johannes Peter (1801–1858) German scientist and zoologist who conducted pioneering efforts in the microscopic study of organisms. He made early classifications of tiny marine animals such as RADIOLARIANS and FORAMINIFERA.

Murray, Sir John (1841–1914) British naturalist and oceanographer, one of the founders of modern oceanography, who is best remembered for his literary classics *The Depths of the Ocean* (1912) and *The Ocean* (1913) and for his mapping and classification of oceanic bottom sediment. He was a member of the famous *Challenger* expedition as a naturalist from 1872 to 1876 and was editor of their 50 volumes of scientific reports. Murray played an important role in fostering the science of oceanography as a distinct entity.

Nansen, Fridtjof (1861–1930) Norwegian oceanographer and explorer who is best remembered for his attempt, beginning in 1883, to the first to reach the North Pole by deliberately allowing his ship, the resiliently built *Fram,* to be entrapped in the ice and then drift with the current toward the pole. The *Fram* drifted to 84°4′ N latitude, then made little or no progress. At this point Nansen's party set out in sledges toward the pole. They did not succeed but did travel to 86°14′, the farthest northern position attained by anyone up to that time. He is also known for his oceanographic work. In 1910 Nansen designed and put to use a special ocean water temperature and salinity sampling bottle, the Nansen bottle. He made observations of directional ice flow versus wind and current, which he shared with VAGN W. EKMAN, who developed equations called the EKMAN SPIRAL to explain the phenomena. Nansen was also awarded the Nobel Prize in peace in 1922 for humanitarian relief work he did for Russian refugees during World War I.

Nares, George Strong (1829–1915) British admiral best known as captain of the famous HMS *Challenger* between 1872 and 1876 during most of its oceanographic expedition. Nares spent much of his life at sea serving on many scientific expeditions that took him from the Mediterranean to the Arctic. He took part in the search for SIR JOHN FRANKLIN in 1852 under Sir Edward Belcher aboard the HMS *Resolute*, and in 1869, while commanding the British ship HMS *Newport*, Nares slipped through the attending flotilla to be the first ship to pass through the newly opened Suez Canal, much to the fury of the French. Many geographic features have been named after him, such as the Nares Plain beneath the SARGASSO SEA in the North Atlantic, the Nares Strait between Greenland and Ellesmere Island, and Nares Mountain and Nares Lake, in Yukon, Canada.

Necho II (c. 658–595 B.C.E.) Pharaoh of Egypt in the 26th dynasty who promoted the first circumnavigation of Africa. He ordered fleets to be built and then crewed by Phoenician sailors. It is said that the voyage took three years. Necho II is also responsible for the first attempt to build a canal, in order to connect the Red and Mediterranean Seas; however, the attempt failed. He also strengthened Egyptian trade with the Greeks by bolstering the Egyptian navy.

Neckham, Alexander (1157–1217) English monk and investigator of science who was the first to introduce to Europe the idea of the mariner's compass, when he made reference to it in his writings around 1190, stating it could help navigators find their way even when weather obscured the celestial sky. He published a scientific encyclopedia, *De natura rerum* (1217), containing a description of a magnetic compass. Neckham also alluded to human underwater activities when he compared the trunk of an elephant to the "tubes used by divers."

Newton, Sir Isaac (1642–1727) English scientist and mathematician who is best known for demonstrating gravitational theory and for his famous three-volume work *Philosophiae naturalis principia mathematica* (1687), in which he describes principles related to the Earth's density, discrepancies in the movements of the Moon, the PRECESSION OF THE EQUINOXES, the increase in gravitation with latitude, and the motion of the tides.

Nordenskjöld, Nils Adolf Erik, baron (1832–1901) Swedish explorer, navigator, and mineralogist who, in 1878–79, was the first to negotiate the NORTHEAST PASSAGE in his ship *Vega*. Earlier he had commanded several expeditions to Arctic regions between 1864 and 1872, during which he explored Spitsbergen, mapped it, and amassed a fine collection of flora and fauna; he also made a journey inland over ice-covered Greenland and later attempted to reach the North Pole, approaching within about 575 miles (925 km), the closest position yet attained. His best-known works are *The Voyage of the Vega* (1881), *Facsimile-Atlas* (1889), and *Periplus* (1897).

Orellana, Francisco de (c. 1511–1546) Spanish explorer who led the first party that completely traversed the Amazon River in South America in 1541, after abandoning the leader of their original military expedition, Gonzalo Pizzaro, in search of food. The search took him unexpectedly down the entire length of the river, which was at first called Rio Santa Maria de la Mar Dulce (the name given by Vicente Pinzón, who thought it was the Ganges), but de Orellana renamed it the Amazon, after the Amazon women in Greek mythology, after an attack by a native tribe of both male and female warriors. He died on a subsequent journey to explore the mouth of the Amazon.

Ortelius, Abraham (1527–1598) Flemish geographer and cartographer, colleague of GERARDUS MERCATOR, who is responsible for producing the first modern atlas, *Theatrum orbis terrarum,* published in 1570. Containing more than 50 maps, Ortelius's atlas was compiled, using the work of over 80 contributing cartographers, and although some were not completely accurate, it was the largest and most reliable collection ever compiled in a single volume up until that time. Navigators were delighted, though they understood that its accuracy could not be relied on completely. In 1575, Ortelius was appointed geographer to Philip II of Spain, and in 1587 he produced his next atlas, which had twice as many maps as the first. He also dealt in rare antiquities and produced other geographic works, such as his *Thesaurus geographicus* (1587).

Owen, Sir Richard (1804–1892) English zoologist and anatomist best known for his numerous contributions to science, many of them in paleontology, and for his opposition to CHARLES DARWIN's theory of evolution. In 1856 he became director of the Natural History Department of the British Museum and remained there for 28 years. Owen, who was vastly interested in fossils, was the person who coined the term *dinosaur.* He also conducted detailed studies of living organisms, dissecting and then carefully recording his findings, especially organisms of the rare type. His most famous work is his five-volume *Descriptive and Illustrated Catalogue of the Physiological Series of Comparative Anatomy* (1833–40). He also wrote *Memoir on the Pearly Nautilus* (1832) and *Anatomy and Physiology of Vertebrates* (1866).

Palmer, Nathaniel Brown (1799–1877) American sea captain and explorer who is thought to be the first explorer to sight the Antarctic Peninsula, in 1820. At the young age of 21, Palmer was in command of the sloop *Hero* and leading his second of what would be seven seal-hunting voyages off Cape Horn in South America, when he caught sight of the peninsula, which he believed to be merely an island. FABIAN G. VON BELLINGSHAUSEN named it Palmer Land while exploring the area during the same year. It was during this trip that Palmer also discovered the South Orkney Islands.

Park, Mungo (1771–1806) Scottish explorer and surgeon who, in 1796, led one of the first European expeditions to investigate the Niger River of western Africa, breaking important ground in Europe's beginning explorations of Africa and its water system. Park discovered that the river turned eastward, not westward, as previously assumed, and after traversing only part of its length he was captured. He escaped, however, and eventually returned to Great Britain in 1797. In 1805 be set out on another expedition to map the course of the Niger; however, his party was attacked in 1806 and Park drowned.

Parry, Sir William Edward (1790–1855) British explorer and naval officer who led three expeditions seeking a NORTHWEST PASSAGE, the first in 1819–20. Approaching the Arctic from the east, Parry sailed though Lancaster Sound, discovering Melville Island and naming Barrow Strait, but he did not succeed in charting the route. Subsequent attempts in 1821–23 and 1824–25 were unsuccessful as well. In 1827 he and his crew tried to reach the North Pole by sledging from Spitsbergen, making it as far north as 82.45° N. latitude, the farthest point on record at the time, but a lack of supplies and human stamina turned them back only 500 miles (800 km) short of their goal. Though unsuccessful in attaining his goal, Parry, through the discovery of the east entrance to the Northwest Passage, confirmed the existence of the route, which ROBERT MCCLURE's party followed to reach Melville Sound over 30 years later on their approach from the west.

Patrick, Ruth (b. 1907) American limnologist (freshwater scientist) known as a pioneering advocate of environmental conservation in her field. To this day, her studies have set the standards of freshwater stream and river management. She is also credited with the invention of the diatometer, a device that precisely measures the presence and quantity of pollution in freshwater.

Peary, Robert Edwin (1856–1920) American naval officer and explorer of great fortitude famous for leading the first expedition to reach the North Pole, in 1909. In the past, Peary usually has received most of the credit for this accomplishment; however, equal credit should also be given to his African-American partner and personal assistant, MATTHEW HENSON, who Peary maintained was "indispensable." Also in the party were four Inuit natives from northern Greenland, Ootah, Seegloo, Oqueah, and Egingwah, whose skill and cultural practices were heavily relied on for survival. Peary's famous trip to the Pole established that, unlike the South Pole, which exists on land, the geographic North Pole resides on a platform of floating ice. It should be said that Fredrick Cook, a former doctor on one of Peary's expeditions, because of racial prejudice against Henson, tried to claim that earlier, in 1908, it

Peary, Robert (Courtesy of NOAA)

was he, not Peary, who first reached the North Pole. Later, however, Cook's claim was discredited.

Peregrinus, Petrus (c. 1220–) French scholar, soldier, and scientist, otherwise known as Peter the Pilgrim, who made important studies of magnetism. In his *Epistola de magnete* (13th century), he described the phenomenon of simple magnetism and was the first to label the ends of a magnet *poles*. His improvement on the compass was to allow the lodestone to pivot inside a graduated directional scale. This improvement held for almost 300 years, until the English physicist William Gilbert published his book *Of Magnets, Magnetic Bodies, and the Great Magnet of the Earth* (16th century).

Peron, François (1775–1810) French naturalist and physicist who laid aside important scientific work in order to serve as head scientist, after the deaths of key colleagues, on a French circumnavigation of the globe commanded by the French explorer NICHOLAS BAUDIN between 1800 and 1804. Peron took a few preliminary deep-sea temperature readings; although the information amounted to little, as a result of it, he became convinced that oceanic research was not being given the attention it warranted. He and Captain L. de Freycinet recorded an account of the voyage, *Voyage de découvertes aux terres australes,* which was printed in 1807.

Piccard, Auguste (1884–1962) Swiss scientist who was a major contributor to the exploration of the oceans during the 1940s. In 1947 he invented the bathyscaphe, a vehicle that could descend deeper into the sea than any other craft to venture there before. The invention of this vehicle made it possible for scientists then and now to conduct studies at crushing depths. His twin brother, Jean-Félix Piccard, was an engineer and chemist, who, along with Auguste, made many high-altitude balloon ascents during the 1930s in order to study cosmic rays.

Piccard, Jacques (b. 1922) Swiss oceanographic engineer, the only son of AUGUSTE PICCARD. He is best known for piloting his father's development, the bathyscaphe *Trieste,* into the MARIANA TRENCH in 1960, accompanied by U.S. Navy Lt. Donald Walsh. This was only one in a series of deep-sea dives sponsored by the U.S. Navy from 1957, during which

Piccard's role was that of scientific consultant. Afterward, while living in Lausanne, Switzerland, Piccard designed and completed the *Auguste Piccard,* a submarine built for tourists, and the *Ben Franklin,* a submarine used for the purpose of studying ocean currents.

See also 1960.

Pinzón, Vicente Yáñez (1463–1514) Spanish navigator and captain of the *Niña,* who sailed in 1492 with CHRISTOPHER COLUMBUS on his voyage to the Americas. In 1499 he voyaged to the Caribbean, the Cape Verde Islands, and Brazil, where he, except for AMERIGO VESPUCCI's claim to be the first, was the first to discover the mouth of the Amazon River, which he mistook for the Ganges, and named it the Rio Santa Maria de la Mar Dulce (later renamed by FRANCISCO DE ORELLANA).

Pliny the Elder (c. 23–79 C.E.) Roman naturalist, born Gaius Plinius Secundus, who compiled information on many scientific and historical issues; however, only his 37-volume encyclopedia *Historia naturalis* (Natural history), in which subjects such as cosmology, astronomy, geography, medicine, zoology, botany, agriculture, ethnology, metallurgy, and even a history of art are covered, remains in existence. Pliny recorded information he gleaned from almost 2,000 books by more than 100 different authors, which he carefully listed as sources. While he was serving as commander of a fleet in the Bay of Naples, Mount Vesuvius erupted above Pompeii. He sailed to shore to investigate, was overcome by noxious fumes, and died.

Ponce de León, Juan (c. 1460–1521) Spanish explorer and soldier who was the first European to reach the coast of Florida, in 1513. Legend has it that Ponce de León was searching for a "fountain of youth" and on arrival during the Easter holidays, he named the new land *Florida* for either the holiday of Pascua Florida or perhaps the abundance of flowers he found growing wild there. He also accompanied CHRISTOPHER COLUMBUS on his second voyage to America in 1493. In 1521 he made a second voyage to Florida with orders to conquer its people but stopped an arrow instead. Mortally wounded, he managed to sail to Cuba, where he died.

Ptolemy, Claudius (c. 90–168 C.E.) Greek mathematician, astronomer, and geographer known for his hypothesis that the Earth was the center of the universe, an accepted belief until the time of Copernicus (16th century), and for his creation of a series of maps that demonstrated curved meridians and a straight center meridian, which he called the prime meridian and, by coincidence, positioned close to the prime meridian. His best-known work is his *Geographical Treatise*, an eight-volume set that holds an atlas of the world as it was known at that time.

Puget, Peter (c. 1762–1822) British naval officer and explorer who accompanied Captain George Vancouver on a circumnavigation of the globe between 1791 and 1795. Their party became the first Europeans to explore western coastal areas of the North Pacific Ocean from today's Washington state as far north as Cook Inlet. Puget Sound, Washington; Cape Puget, Alaska; and Puget Island in the Columbia River are all named for him.

Pytheas (c. 350 B.C.E.) Greek geographer and explorer who made an early reference to the tides' being somehow affected by the Moon and realized that the North Star was not situated directly above the North Pole. He was an exceptional navigator, considering the era in which he lived, and confounded the Carthaginian navy by slipping through their blockades in order to explore coastal areas of Europe.

Queirós, Pedro Fernandes de (1565–1615) Portuguese explorer who, while sailing in the service of Spain, took part in the discovery of a group of 10 highly fertile volcanic islands called the Marquesas Islands. De Queirós (spelled *de Quirós* in Spanish) believed in the existence of a great continent in the South Pacific, but throughout his explorations he discovered only small Pacific islands.

Ray, John (1627–1705) English naturalist and influential philosopher and theologian who first initiated a taxonomic classification system for organisms. His work led to the botanical classification work done by the famous Swedish botanist CAROLUS LINNAEUS. In 1648 Ray graduated from the University of Cambridge; he then worked for 13 years on a fellowship, cataloging the local flora.

Between 1661 and 1671 he traveled throughout Europe collecting flora, fauna, and minerals. Ray was the first to define the term *species* in the modern sense. Among his works are *Catalogue of Cambridge Plants* (1661) and *Methodus plantarum nova* (1862). One of his most influential works was the three-volume *Historia plantarum* (1686–1704). In 1844 the Ray Society of London was established in his honor for the purpose of publishing scientific works.

Rebikoff, Dimitri (b. 1921) French engineer and pilot who made great contributions to the development of oceanographic techniques and underwater capabilities. In the 1940s and 1950s, he was a pioneer in the innovation of underwater diving, particularly underwater photography. In 1952 he invented the *Torpille,* the world's first underwater scooter, and a year later he completed the cigar-shaped *Pegasus/Remora,* a self-propelled underwater camera and personnel vehicle. Rebikoff also invented a highly precise correction lens for underwater photogrammetry (the science of using photographs to make reliable measurements) and a color and temperature measuring meter. In 1947 he invented the portable flash. As technology blossomed, Rebikoff was able to develop mechanisms such as high-speed underwater film cameras used by the movie industry, the U.S. Navy, and oil research companies. In 1980 he established a nonprofit organization, the Institute of Marine Technology, in Fort Lauderdale, Florida.

Rennell, James (1742–1830) British oceanographer, geographer, and cartographer who pioneered systematic studies of ocean flow, believing that surface currents are a product of the wind. From 1764 to 1777 he served as surveyor general of Bengal, India, and in 1779 he published *A Bengal Atlas.* In 1783 he developed the first accurate map of India. Rennell supplied an appendix and geographical illustrations for the Scottish explorer and surgeon MUNGO PARK's *Travels in Africa* (1799).

Revelle, Roger (1909–1991) American oceanographer and sociologist who, while working for the Scripps Institution of Oceanography, conducted surveys of seafloor topographic features and ocean SEDIMENT in the Pacific Ocean. He was a major contributor to the theory of SEAFLOOR SPREADING.

Rondelet, Guillaume (1507–1566) French physician and zoologist who made significant contributions to the knowledge of marine biology. In 1555, while teaching medicine at the University of Montpellier, France, he published his celebrated *Universae aquatilium historiae* (Complete history of fish), within which he identified nearly 300 species of marine organisms with accompanying illustrations.

Roosevelt, Nicholas (1767–1854) American engineer, great-granduncle of President Theodore Roosevelt, who was a pioneer in steamship navigation. One of his early attempts at ship design occurred in 1782, when he built a model boat with side-paddle wheels turned by springs. Later, in Belleville, New Jersey, Roosevelt established a foundry where he built a steamship that unfortunately was not successful. In 1811 he worked with ROBERT FULTON in designing and building the *New Orleans,* the first well-constructed steamer to navigate the Ohio and Mississippi Rivers successfully.

Ross, Sir James Clark (1800–1862) British polar explorer who is best known for discovering the magnetic North Pole in 1831 while on an expedition with his uncle, SIR JOHN ROSS. Between 1819 and 1827, during four expeditions led by WILLIAM E. PARRY, he helped explore Lancaster Sound, Melville Island, and Prince Regent Inlet. In 1839 he set out in search of the magnetic South Pole, and although he did not secure its location, he was the first to sight Victoria Land and subsequently traveled farther south than all previous explorers. During this voyage he made the first extensive deep-sea soundings and temperature readings of the southern waters and made dredgings down to 2,400 feet (732 m), though documentation of specimens collected did not survive. The Ross Sea and Ross Island are named in his honor. Ross wrote *A Voyage of Discovery and Research to Southern and Antarctic Regions* (1847).

Ross, Sir John (1777–1856) British polar explorer and uncle of Sir JAMES CLARK ROSS, who carried out pioneering efforts in the field of marine research. In Baffin Bay, Canada, 1817–18, while conducting his first expedition in search of the NORTHWEST PASSAGE, he used a deep-sea clam dredge to haul to the surface from depths as far as 5,900 feet (1,798 m) living organisms, disproving the theory of EDWARD FORBES

(1815–54), who insisted that because of the lack of light and oxygen there could be no life below the depth of 1,600 feet (488 m). Though he made wonderful breakthroughs in the science of marine biology, Ross never found the Northwest Passage and mistakenly claimed it did not exist.

Rossby, Carl-Gustav Arvid (1898–1957) Swedish meteorologist best known for his discovery of what are now called Rossby waves, an explanation of how the air flows within the JET STREAM, and the Rossby equation, a determination of the speed at which the flow develops. He also made studies of turbulence and deep-circulation routes within the oceans.

Sars, George Ossian (1837–1927) Norwegian marine biologist, son of MICHAEL SARS, best known for his work on Crustacea. He was the leading expert on crustacean species in his time and was called on to examine all crustaceans takes up by the *Challenger* dredgings. He also conducted studies of the reproductive processes of the codfish and concluded, among other things, that their eggs floated, rather than sank, as believed. This information was important in documenting the migratory patterns of important food fishes.

Sars, Michael (1805–1869) Norwegian marine biologist who made significant contributions to the knowledge of deep-sea organisms. He successfully set out to prove the inaccuracy of EDWARD FORBES's theory that no life existed in the ocean below 1,600 feet (488 m) and provided documentation of living things dredged up from deep Norwegian fjords, such as the first living stalked crinoid (a class of echinoderm) ever found. His findings aroused greater curiosity about the deep, and expeditions such as the famous *Challenger* expedition were results of his work. He wrote *Contributions to the Natural History of Marine Animals* (1829) and *Descriptions and Observations* (1835). In 1997, the Sars International Center for Marine Molecular Biology was founded in Bergen, Norway.

Schmidt, Ernst Johannes (1877–1933) Danish biologist and director of the Carlsberg Physiological Laboratory, Copenhagen, who conducted detailed research on the European eel in order to understand their life cycle better and determine where their breeding grounds might be. In 1904 Schmidt concluded that their principal breeding point must be the SARGASSO SEA, for he found the very young eels living in those waters and from that point outward the eels encountered were more mature. After World War I (1914–18), using the sister ships *Dana I* and *Dana II,* Schmidt led a series of expeditions that continued throughout the 1920s. The great collection of marine organisms resulting from those explorations laid the groundwork for the *Dana Reports,* which are still useful today.

Scoresby, William (1789–1857) English scientist and Arctic explorer who began life as a whaler, participating in many whaling voyages with his father in the Arctic waters near Greenland. In 1810–22 he explored these regions and made many deep-sea and surface observations, including surveying hundreds of miles of East Greenland coastline. In 1820 he published *The Arctic Regions,* the first scientific account of the waters and lands of the Arctic. After 1822 he retired from ocean exploration to enter the clergy; however, in 1856, while voyaging to Australia, he conducted studies on the magnetism of Earth.

Scott, Robert Falcon (1868–1912) British Antarctic explorer best known for leading the second team to reach the South Pole, trailing behind ROALD AMUNDSEN by only one month. Scott and his party of four other men were crestfallen when in January 1912 they arrived at the pole to find Amundsen's markers already there. On the return trip, one after another, their members died; their diaries were found less than a year later by search parties. Before the trip that ended his life, Scott belonged to the British navy and commanded the HMS *Discovery* during the British National Antarctic Expedition of 1901–03, during which he explored Victoria Land (Antarctica) and at one point reached 82° S, a record penetration southward up until that time.

Shackelton, Sir Ernest (1872–1922) British explorer who took part in many expeditions to reach the South Pole. His most famous and fantastic endeavor was an attempt to be the first to lead a party of men over the Antarctic continent. On the British Imperial Trans-Antarctic Expedition, Shackelton set out in 1914 aboard the ship *Endurance* together with 28 crew members and scores of sledge dogs with a plan to cross from a base camp in the Weddell Sea, over the pole, and arrive at the Ross Sea's McMurdo Sound 2,000 miles (3,200 km) away. The expedition had just begun when the *Endurance* was crushed in pack ice. The party drifted for months on floes, living mainly off seal meat cooked over seal blubber flames, until finally they landed on Elephant Island, the first dry land they had seen for over 400 days. Leaving most of the men on the desolate island, Shackelton and a handful of others crossed the wild waters of the Drake Passage in a pair of open lifeboats, finally to land on South Georgia Island 800 miles (1,287 km) away to seek help from the Norwegian whaling station based there. Shackelton returned to Elephant Island, as promised, and rescued the remainder of his men. Remarkably, no member of his team perished during the ordeal that had placed them at the mercy of the Antarctic for more than two years. Shackelton wrote *The Heart of Antarctica* (1909) and *South* (1919) describing his voyages. Another good account of his famous trans-Antarctic attempt is Alfred Lansing's *Endurance* (1959).

Shepard, Francis Parker (1897–1985) American marine geologist who made great contributions to the understanding of marine geology through his teaching and published writings. It is said that he was the first to take up marine geology as his main scientific study. Shepard was a professor of geology at the University of Illinois from 1939 to 1946. Between 1937 and 1942 he spent his summers working with the Scripps Institution of Oceanography, and from 1942 to 1945 he was the principal geologist for the University of California Division of War Research. His chief works are *Submarine Geology* (1948), *The Earth beneath the Sea* (1959), and *Our Changing Coastlines* (1971). Throughout his career, Shepard received numerous awards and medals for his outstanding work in the field of marine science.

Charles D. Sigsbee piano wire sounding machine
(Courtesy of NOAA)

Sigsbee, Charles Dwight (1845–1923) American naval officer with an active interest in oceanography. During his long career with the U.S. Navy, between 1873 and 1888 he served in the Hydrographic Office and the Coast Survey, in which he invented numerous deep-sea sampling and sounding devices. During peacetime, ships under his command conducted the first real deep-water studies of the Gulf of Mexico and discovered the Sigsbee Deep, the deepest point in the Gulf of Mexico. In regard to marine science, Sigsbee wrote *Deep-Sea Sounding and Dredging* (1880), dealing with the findings and techniques used for surveying the deep.

Siple, Paul Allman (1908–1968) American Antarctic explorer chosen by Adm. Richard E. Byrd to accompany him on his first Antarctic expedition of 1928. Between 1933 and 1935, Siple headed the biology team for Byrd's second expedition, and between 1939 and 1941, he was the geographer on a U.S. expedition. From 1946 to 1957, Siple participated in many major expeditions as a member of the U.S. Department of War. It has been said that he spent more time in the Antarctic waters than any other human, and he is given credit for the formulation of the wind chill factor. Siple wrote *A Boy Scout with Byrd* (1931), *Exploring at Home* (1932), and *90 Degrees South—the Story of the American South Pole Conquest* (1959).

Sperry, Elmer Ambrose (1860–1930) American scientist and inventor whose best-known development are the gyroscope-guided automatic pilots used by ships for oceanic navigation; his invention is also used in aircraft and some spacecraft. Because of the amount of steel used in the construction of ships at the turn of the 20th century, magnetic compasses were becoming unreliable. In 1911, Sperry combined mechanical and electrical elements to invent a gyroscopic compass, which solved the problem.

Steller, George Wilhelm (1709–1746) German naturalist who traveled with VITUS BERING on his voyage to Alaska in 1741, during which he studied many animal species. He was the first to conduct detailed analysis of a large species of sea lion, later named the Steller sea lion. He also discovered a seabird, the spectacled cormorant, while shipwrecked on what was later named Bering Island in the western Aleutian chain; the bird today is now extinct. The blue Steller's jay is also named for him.

Stevens, John Cox (1749–1838) American inventor who helped pioneer the steam engine. Though ROBERT FULTON's *Clermont* was the first successful steam-powered riverboat, Stevens's

Phoenix, which he designed and completed in 1809, was the world's first oceangoing steamship. He operated the first American steam railway locomotive on his grounds in 1825.

Stevenson, Robert (1772–1850) Scottish engineer who between 1811 and 1833 built 23 lighthouses along the coast of Scotland and Britain and introduced in their design a flashing light with which to guide ships. He succeeded his stepfather, Thomas Smith, as engineer to the Lighthouse Board of Glasgow, Scotland, and served as such for 47 years, so launching a family tradition known as the dynasty of the "Lighthouse Stevensons." Between 1790 and 1940, eight family members took up the proud trade, erecting 97 lighthouses of their own design along the coast of Scotland. Robert Stevenson was grandfather to the famous author Robert Louis Stevenson.

Stimpson, William (1832–1872) American naturalist and malacologist (researcher of mollusks) who greatly expanded the knowledge of marine organisms as naturalist on the North Pacific Exploring Expedition of 1853–56. The expedition, led first by Captain Cadwallader Ringgold and later by Captain John Rodgers, was commissioned by the U.S. Congress to survey commercially traveled routes between the United States and China. Stimpson took back one of the largest collections of specimens produced by any voyage, some of which were added to the collections of the Smithsonian Institution. The eel *Bathycongrus stimpsoni* and the goby fish *Sicydium stimpsoni* are named for him.

Stommel, Henry Melson (1920–1992) American oceanographer who pioneered studies of deep-sea circulation and in the 1960s invented a new process to derive the speed and salinity of bottom currents. His studies of the GULF STREAM contributed greatly to our current knowledge. Stommel inspired and collaborated with dozens of scientists and published hundreds of scientific papers. His many writings include *Science of the Seven Seas* (1945), *The Gulf Stream* (1966), *Lost Islands* (1984), and *Introduction to the Coriolis Force* (with Dennis Moore, 1989). *The Collected Works of Henry Stommel* were published posthumously in three volumes in 1995.

Suess, Eduard (1831–1914) Austrian geologist who wrote *Das Antlitz der Erde* (The Face of the Earth, 1883–1909), in which he proposed a theory that the southern continents of Earth were once one great supercontinent, which he named GONDWANALAND; his theory helped steer the scientific public toward the theory of continental drift. Between 1857 and 1901, he was a professor of geology at Vienna University.

Sverdrup, Harald Ulrik (1888–1957) Norwegian oceanographer who spent seven years studying the dynamics of the Arctic, beginning in 1918, first with ROALD AMUNDSEN's party and then aboard the *Maud* as he tried to duplicate the journey made by the *Fram*. Sverdrup took charge of scientific work aboard the submarine *Nautilus* in 1931. In 1936 he became director of Scripps Institution of Oceanography in the United States, where he founded the Marine Life Research Program, ongoing today, and initiated the first systematic oceanography course offered in America. He wrote more than 50 papers documenting his oceanographic research during his explorations of the Arctic.

Swallow, John (1923–1994) British oceanographer and lifelong friend of HENRY STOMMEL who conducted studies of ocean circulation at all depths and in 1956 invented the Swallow float, a ballasted device that could be set at controlled depths and free-float, allowing a ship to monitor its position continuously by using acoustics. Between 1960 and 1994 he received numerous medals and honors in recognition of his work, which radically expanded the knowledge of the dynamics of the world's oceans.

Tasman, Abel Janszoon (c. 1603–1659) Dutch navigator and explorer who in 1642 set out to explore Australia. Though others had gone before him, he succeeded in circumnavigating the entire landmass and explored an island he named Van Diemen's Land, after Anton van Diemen, the governor-general of the Dutch East India Company, for whom he worked. Van Diemen's Land is now known as Tasmania, named for Tasman, along with the Tasman Sea. Much of the southern regions was uncharted, and Tasman mapped the areas around the New Hebrides, the islands of Tonga and the Fiji archipelago, New

Guinea, and the Solomon Islands. In 1644 he voyaged again and explored the Torres Strait and the Gulf of Carpentaria.

Thomson, Charles Wyville (1830–1882) Scottish naturalist and oceanographer who, in 1872 with the help of the Royal Society of London, headed the first voyage launched specifically to study the world's deep seas. Aboard the HMS *Challenger,* Captain Thomson spent three years covering nearly 70,000 miles (110,000 km) and completing 362 deep-sea soundings across the Atlantic Ocean, Pacific Ocean, and the Indian Ocean. Thomson wrote *The Depths of the Oceans* (1872), in which he describes events and findings of his famous deep-sea researching. He was educated at the University of Edinburgh and during his career he served as professor of natural history at Cork, Belfast (1854–68), and Edinburgh, Scotland (1870–82), and lectured at King's College, Aberdeen. His remarkable work was crucial in the development of modern oceanography.

Thomson, William (Lord Kelvin) (1824–1907) British scientist who developed a machine that could predict the tides and made improvements on navigational compasses. He also developed what was called the mirror-galvanometer, a device that made possible the first successful sustained telegraph transmissions in transatlantic submarine cable. In science, he conducted research in thermodynamics and devised the Kelvin temperature scales named for him. In 1851 Thomson was elected to the Royal Society of London, and in 1866, for his contribution in making the transatlantic cable a success, Thomson was knighted.

Varen, Bernhard (1622–c. 1650) Also known as Varenius Bernhardus, a German physician and geographer who pioneered methods in modern geography. In his 1650 book *Geographia generalis,* he recorded the first account of wind-driven ocean currents, which led scholars in a new direction in understanding the motions of the sea. He also described his studies on tides and their association with the Moon.

Vening-Meinesz, Felix Andries (1887–1966) Dutch geodesist known for enhancing knowledge of the measurement of gravity and for uncovering the presence of gravitational anomalies along the seafloor. He developed a three-pendulum apparatus to be used aboard a submarine while underwater for measuring gravity on the seabed. Between 1923 and 1927, Vening-Meinesz discovered along the island chain of Indonesia and parts of the western Pacific Ocean's abyssal trenches some of the largest low-gravity belts ever recorded. His discoveries opened new and favorable ground in the advancing theory of plate tectonics. Vening-Meinesz was president of the International Association of Geodesy between 1933 and 1955 and of the International Union of Geodesy and Geophysics from 1948 to 1951.

Verrazano, Giovanni da (c. 1480–c. 1528) Italian navigator who sailed for the French maritime service in search of a sea route to China. In 1524 he landed on the east coast of America and explored the area from modern North Carolina to Nova Scotia and was the first European to discover New York and Narragansett Bays. The Verrazano Narrows Bridge in New York Harbor between Brooklyn and Staten Island is named in his honor.

Vespucci, Amerigo (1454–1512) Italian navigator who claimed to the be the first to reach the shores of the New World in 1497 and in whose honor America was named. It was the German cartographer Martin Waldseemüller who proposed the name *America*, modified from *Amerigo*, and published a planisphere in 1516 exhibiting the new name, which eventually was used for both Western Hemisphere continents. In 1499 Vespucci sailed with the Spanish expedition leader Alonso de Ojeda to explore the coast of South America. The two split up, and afterward he discovered the Amazon River, though VICENTE YÁÑEZ PINZÓN is usually regarded as its official discoverer. As a navigator, Vespucci was remarkable because, rather than rely on dead reckoning, as did other navigators, he devised an accurate system to work out longitude. His estimate of the circumference of the equator fell short only 50 miles (80 km) of the true measurement.

Villepreux-Power, Jeanne (1794–1871) French marine biologist famous as the first to invent (1832) and use indoor aquariums in order to conduct aquatic experiments in a controlled environment. In 1858 Professor RICHARD OWEN recognized her as the "mother of aquariophily," and for 10 years (1832–42) she was the single female member of the London Zoological Society. She completed several works, some of which can be found in the major European natural history libraries, and along with her legacies within scientific literature her name has been immortalized by the designation of an important crater on Venus, the Villepreux-Power crater.

Vine, Fred (1939–1988) British geologist who in the early 1960s, along with his associate Drummond Matthews, made magnetometor surveys of the Atlantic Ocean's midoceanic ridge and discovered MAGNETIC ANOMALIES distributed evenly on both sides of the ridge. Furthermore, they found the same symmetrical anomalies along different areas of the ridge and recognized that the patterns of the anomalies in the sea matched known patterns of global magnetic reversals. Their findings were instrumental in proving the theory of plate tectonics. Vine and Matthews showed that these seafloor magnetic anomalies on each side of midocean ridges could be used as a record of the action of SEAFLOOR SPREADING and studied the relation of the spreading to the continental drift theory.

See also MENARD, HENRY WILLIAM; WEGENER, ALFRED LOTHAR.

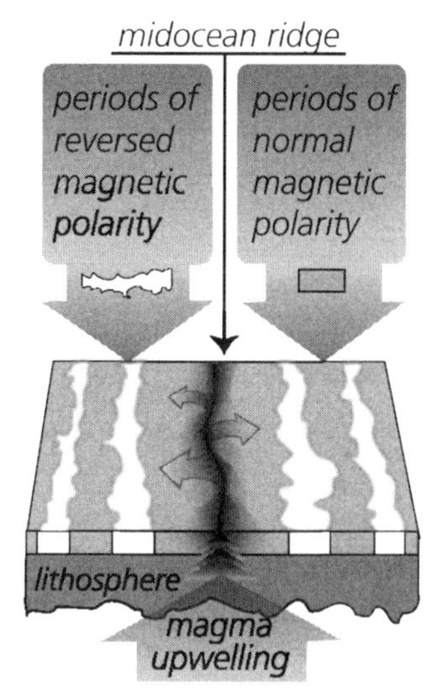

Magnetic anomalies discovered by Fred Vine

Waghenaer, Lucas Janszoon (c. 1534–1605) Dutch cartographer and sea pilot who in 1582 began gathering data for the creation of pilot's charts for Western and Northern Europe. The finished work, *Mirror of Seafaring*, published in 1584, contained fairly accurate sea and coastal charts of the region from Holland to Spain and the Baltic and North Seas. Waghenaer's work caught the appreciative eye of the lord high admiral of England, who had it translated into English and republished in 1588 under the new name *The Mariner's Mirror.* The charts grew in popularity to the extent that the name *Waggoner,* a derivation from *Waghenaer,* came to be a universal term in English for sea charts of any kind.

Wallace, Alfred Russel (1823–1913) British self-educated naturalist who learned his trade through diligence in the field and passion for the subject. He discovered the evolutionary theory of natural selection independently of CHARLES DARWIN; he later sent his records to Darwin, who presented them along with his own in 1858. Wallace wrote a book on natural history, *The Malay Archipelago* (1869), which was widely popular and reissued several times over the years. He also wrote a *Species Notebook*, which divided animals according to their geographic habitats and their related species; he based his conclusions on his observations of the flora and fauna in the South American and Malaysian regions. The concept of an imaginary boundary between species came to be known as the WALLACE LINE.

Weddell, James (1787–1834) English navigator, seal hunter, and explorer who between 1815 and 1824 made significant voyages into the Southern Ocean around Antarctica. On one such trip he explored the South Shetland Islands and named the South Orkney Islands. During his final voyage he made a record penetration of the south at latitude 74° before the ice turned him back. Weddell wrote the popular *A Voyage toward the South Pole* (1822). In Antarctica the Weddell seal and WEDDELL SEA were both named in his honor.

Wegener, Alfred Lothar (1880–1930) German meteorologist and physicist famous for proposing the once-controversial theory of CONTINENTAL DRIFT, in 1912. He observed the paths of continental coastlines and insisted they were like pieces of a disassembled jigsaw puzzle and that they had, over millennia, drifted apart. However, though he tried, Wegener could not thoroughly explain how this drift might have happened until the work of scientists such as FRED VINE proved the phenomenon of SEAFLOOR SPREADING and the existence of PLATE TECTONICS.

See also MENARD, HENRY WILLIAM.

Wilkes, Charles (1798–1877) American naval officer and explorer who joined the U.S. Navy in 1818 and in 1830 became head of the Department of Maps and Charts. He lobbied Congress until, in 1838, they finally granted him a six-vessel expedition to explore parts of the coast of North America and the south polar continent, which Wilkes named Antarctica. Among the expedition's accomplishments were a circumnavigation of the globe, the charting of hundreds of Pacific islands, and the extensive research done on marine organisms, namely, Crustacea. The voyage ended in 1842. Wilkes wrote *Narrative* *of the United States Exploring Expedition,* in five volumes (1844), and edited more than 200 books on scientific data.

Willoughby, Sir Hugh (d. c. 1554) English navigator who, in 1553, set sail northward from England in search of the NORTHEAST PASSAGE to India. In command of three ships, the *Bona Esperanza* (Willoughby's), the *Bona Confidentia,* and the *Edward Bonaventure,* they sailed only as far as the Lofoten Islands off Norway before weather separated their party. With his two remaining ships, Willoughby sailed into the Barents Sea and eventually landed on the coast of Scandina at the mouth of the Arzina River, where they attempted, without success, to wait out the winter months. Years later, their bodies, ships, and journals—with the last entry dated 1554—were found by the captain of the *Edward Bonaventure,* RICHARD CHANCELLOR.

Wilkes, Charles (Courtesy of NOAA)

Wilson, John Tuzo (1908–1993) Canadian professor of geophysics at the University of Toronto who is best known for his 1965 explanations of SEAFLOOR SPREADING. He graduated from the same university with a degree in geophysics, later earned his doctorate, and is known for coining the term *plates.* Wilson's influential studies of magnetic anomalies in the ocean floor, transform faults, and seismic activities of the ocean crust helped prove the theory of CONTINENTAL DRIFT.

Wrangell, Ferdinand Petrovich, baron von (1796–1870) Russian naval officer and diplomat known for his explorations of the Arctic, between 1820 and 1824, during which he conducted extensive observations and surveys. One year later he set sail on a circumnavigation of the globe that lasted from 1825 to 1827. In 1829 he became the first governor of Russian lands in Alaska; he remained in that position for six years and later strongly opposed the sale of Alaska to the United States in 1867. Wrangell Island, Alaska, is named in his honor.

Wüst, Georg Adolf Otto (1890–1977) German oceanographer best known for his association with the famous German Atlantic Expedition of 1925–27, better known as the Meteor Expedition, named for their research ship, the RV *Meteor.* Using complex equipment, their goal was to study the characteristics of the Atlantic Ocean, reporting on topographic, and biological conditions, salinity, and temperature from the surface to the ocean floor. Wüst is known for his theory on circulation, which details a four-layered system and is still considered valid today.

Yonge, Charles Maurice (1899–1986) English marine biologist known for his studies in malacology (mollusks) and the physical and ecological characteristics of coral reefs, in particular, the Great Barrier Reef off Queensland, Australia. In 1983 the Malacological Society of London established the Charles Maurice Yonge Award, offered for outstanding scientific contributions submitted in writing and published in the *Journal of Molluskan Studies* on the subject of Bivalvia. He wrote *Queer Fish: Essays on Marine Science and Other Aspects of Biology* (1928), *A Year on the Great Barrier Reef: the Story of Corals and of the Greatest of their Creations* (1930), and *British Marine Life* (1944).

Young, Roger Arliner (1899–1964) Remembered as the first African-American woman to earn a doctorate in zoology. She entered Howard University, Washington, D.C., in 1916 and graduated in 1921. In 1924 she entered the University of Chicago as a graduate student and earned a master's degree in 1926. Forced to care for an invalid mother and balance teaching with research, she was prone to bouts of depression. However, despite daunting obstacles, such as failing her qualifying exams for Chicago's Ph.D. program, she was very talented and completed her doctorate at the University of Pennsylvania in 1940. Young worked for the Woods Hole Marine Biological Laboratory and in 1929 was acting head of zoology at Howard University.

SECTION THREE
CHRONOLOGY

4500 B.C.E. • In Mesopotamia, archaeologists unearthed shells of cowrie, a marine gastropod that lives in warm waters on the seafloor, evidence suggesting not just the existence of trade with coastal communities, but also diving to depths in the sea during this era by ancient humans.

c. 4000 B.C.E. • The Egyptians developed shipbuilding and skills to pilot the sea. The earliest ship of which we have definitive knowledge is the funeral ship of Pharaoh Cheops, which was not intended as a seagoing vessel.

c. 3000 B.C.E. • The Egyptians invented the sail, by rigging reed boats with a square sheet set to a central mast. Evidence of this practice is an early model of a ship unearthed in Mesopotamia that had an aperture to hold a mast and holes for ropes.

c. 2500 B.C.E. • Early Greeks practiced diving techniques, carrying sponges to the surface. In both the *Illiad* and the *Odyssey* the act of diving, by means of holding a heavy rock and pouring oil into one's ears, is mentioned as a means of retrieving sponges.

1000–600 B.C.E. • The Phoenicians explored the Mediterranean Sea and reached England.

600 B.C.E. • According to Greek history, the Egyptian pharaoh NECHO II ordered a group of Phoenicians on an expedition to circumnavigate the continent of Africa by maintaining sight of the coastline; the voyage was said to have taken three years.

c. 450 B.C.E. • The Greek historian Herodotus used the term *Atlantic* for the first time to describe the seas to the west and wrote that observations were being made of the regular tides in the Persian Gulf and of the deposition of sediment in the Nile Delta.

c. 350 B.C.E. • The Greek philosopher and scientist ARISTOTLE described in his book *Historia animalium* how sharks generated live young and concluded that dolphins were not fish. He also mentioned a kettle style of dive bell used to assist sponge fishers while harvesting the sea bottom. Aristotle gathered organisms from the Aegean Sea and described and named 24 species of crustaceans and annelid worms, 40 species of mollusks and echinoderms, and 116 species of fish.

332 B.C.E. • Alexander the Great employed divers and a dive bell, reportedly used by him on occasion, to assist during his famous siege of Tyre (Lebanon) in order to eradicate the waters of obstacles posing a threat to shipping.

276–192 B.C.E. • Using trigonometry, the Greek mathematician, astronomer, geographer, and poet ERATOSTHENES correctly determined the circumference of the Earth.

c. 221–206 B.C.E. • In China, the world's first compass was developed by balancing a ladle-shaped piece of lodestone (magnetite) on a bronze plate. This early version was not used for navigation of the seas but for divination, such as determining where to place furniture in the house or where to lay a cornerstone for the erection of a new dwelling.

early lodestone compass

lodestone

100 B.C.E. • Throughout the seaports of the Mediterranean, the practice of salvage diving became so common that a rate of pay was established.

4 B.C.E.–65 C.E. • The Roman philosopher, statesman, and dramatist Seneca put forward his hypothesis on the HYDROLOGICAL CYCLE of water.

150 C.E. • The Greek astronomer and mathematician CLAUDIUS PTOLEMY assembled the *Roman World Map,* which included basic lines of LATITUDE and LONGITUDE.

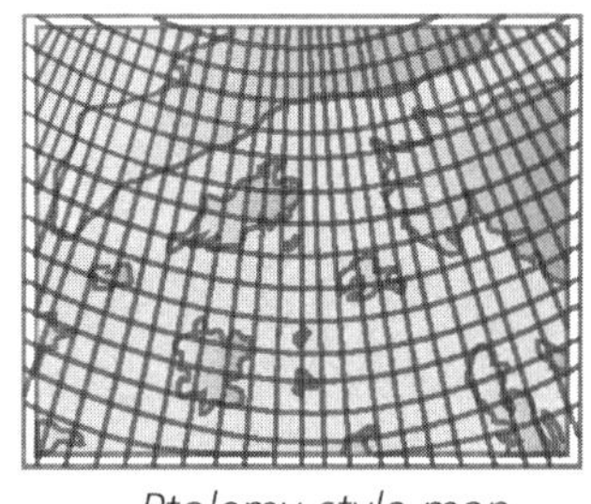

Ptolemy-style map

Ptolemy map

673–735 • The English monk known as the Venerable Bede, who wrote on practically every area of knowledge, described the lunar influences on tidal cycles and recognized the monthly tidal variations.

981 • Banished from his home in Iceland for homicide, the Norwegian explorer ERIC THE RED organized an expedition, set sail, and in 981 discovered Greenland.

995 • The Norwegian LEIF ERICSON (Leiv Eiriksson), or Leif the Lucky, son of Eric the Red, voyaged to the North American coast of Newfoundland and wintered in Vinland, a settlement soon abandoned because of repeated *Skraeling* (Viking name for North American natives) attacks. Vinland's precise location has never been determined.

1080 • Shen Gua of China set down the first precise written description of a compass using a magnetized needle and a clear description of the phenomenon of magnetic variation. The magnetic compass made travel across open ocean possible, and its discovery was a significant event in furthering the civilization of the world.

1180 • The Arabians invented the rudder and tiller as means of steering a ship, an improvement on the existing means of steerage, which was typically done using an oar or a "steer board."

1420s–1430s • The Portuguese HENRY THE NAVIGATOR directed major voyages in European waters that produced several important discoveries, including Madeira, the Azores, and Cape Verde Islands.

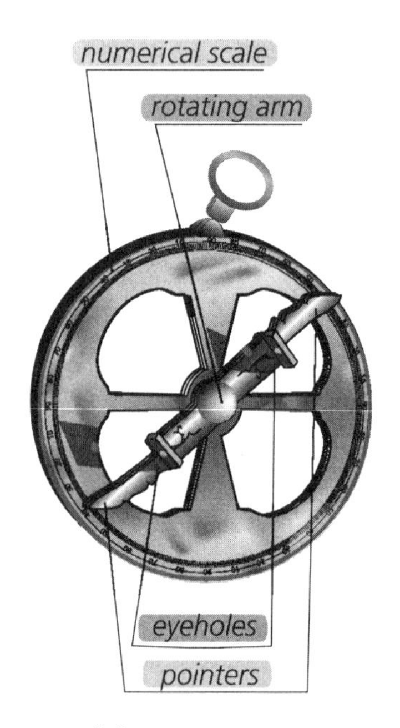

astrolabe

1452–1519 • The Italian artist, architect, engineer, and scientist LEONARDO DA VINCI observed and recorded the motions of waves and currents. From his study of fossils, he deduced that in the past, sea levels must have been higher. He also made an early discovery of how sound propagates through water, stating at one point that the noises created by the movement of ships can travel for great distances through water. Da Vinci designed the first modern "snorkel" device, along with dive fins for the hands and feet.

c. 1470 • Europe introduced the mariner's astrolabe, a version of the astronomer's astrolabe, as a means of navigation for seafarers. The mariner's astrolabe was simply a ring marked in degrees for measuring celestial altitudes, latitudes of lands, and locations at sea.

1492 • The Spanish-Italian navigator CHRISTOPHER COLUMBUS rediscovered the "New World" of the Americas by way of the West Indies.

1498 • The Portuguese explorer VASCO DE GAMA, by rounding the Cape of Good Hope, Africa, successfully reached India and returned to Portugal, discovering for Europe a vital new trade route and thus eliminating the expensive tolls suffered by using the land routes from Italy through Egypt or Arabia or Persia.

1513–1518 • The Spanish explorer VASCO DE BALBOA, after hearing reports of the existence of a new large ocean, headed an expedition to search for it. After spotting the ocean from a Panamanian mountaintop, his party crossed the Isthmus of Panama and sailed triumphantly into the untried waters of the Pacific Ocean, which Balboa initially named Mar del Sur, or South Sea.

1519–1522 • The Portuguese-born Spanish explorer and navigator FERDINAND MAGELLAN set off to circumnavigate the world in 1519; this voyage was completed in 1522 by SEBASTIAN DEL CANO after Magellan's death in the Philippines.

1551 • The French biologist PIERRE BELON published his book *Natural History of Strange Marine Life,* which include recognition of cetaceans as air-breathers.

1589 • The first dredging vessel, a ship capable of removing debris such as silt and seaweed from trafficked waterways, was developed in the Netherlands.

1593 • The Italian astronomer Galileo Galilei invented the first known air thermometer, then called a thermoscope; however, its accuracy was considered only fair. Thermometers are a vital component when testing the sea for temperature differences.

See also 1641; 1709; 1714.

thermoscope

1606 • The Dutch explorer William Jansz became the first European to explore the Australian coast.

1624 • The Dutch scientist Cornelius van Drebble invented the first workable submarine, which was made from a wooden rowboat covered with waterproof hides. Leather washers were used to seal oar holes in the sides, and though the ship leaked copious amounts of water, it was nevertheless propelled along the Thames River in London at a depth of about 13 feet (4 m).

1641 • The first accurate thermometer using alcohol was invented by Grand Duke Ferdinand II, a Tuscan.
See also 1593; 1709; 1714.

1642 • The Dutch navigator ABEL J. TASMAN discovered New Zealand, which he initially named Staten Land in the belief that it was part of South America's southeastern coast called Staten Landt. In 1645 the Dutch cartographer Joannes Blaeu renamed it Zeelandia Nova, or Nieuw Zeeland.

1665 • The Royal Society of London issued orders to all seafarers to collect information about tides, currents, and water depths wherever they went.

1674 • The English natural philosopher ROBERT BOYLE carried out pioneering oceanographic measurements on temperature, salinity, pressure, and depth. He played an integral role in the founding of the ROYAL SOCIETY OF LONDON, a British organization committed to progress in science.

1675 • The Royal Observatory at Greenwich, England, set the line of longitude at 0°, the Greenwich Median. The observatory was founded in this same year.

1681 • The French priest Abbé Jean de Hautefeuille published his *Art of Breathing Underwater,* in which he described why humans cannot breathe at the normal atmospheric pressure while existing at depths beneath the sea.

1687 • The English mathematician and physicist SIR ISAAC NEWTON published his *Philosophiae naturalis principia mathematica* (Mathematical principles of natural philosophy), within which he explains, among many other phenomena, the first mathematical formulation of sound propagation through water.

1690 • The English astronomer Sir Edmund Halley developed a type of dive bell that can be restocked by the use of weighted, air-filled barrels sent down to the work area.

1698–1700 • The English astronomer Sir Edmund Halley sailed as far as 52° S latitude into the Atlantic Ocean to study variations in the magnetic compass. This was most likely the first solely science-oriented voyage made by humans.

1700 • The English mathematician and physicist SIR ISAAC NEWTON discovered the theory behind the sextant but kept it to himself except in a letter sent to a colleague, the English astronomer Sir Edmund Halley.

1709 • The German physicist Gabriel D. Fahrenheit (1686–1736) invented an alcohol thermometer.
See also 1593; 1641; 1714.

1712 • The English inventor Thomas Newcomen built the first steam-powered piston engine. A more accurate name for his device was the atmospheric steam engine. Steam would fill the cylinder of the pumping engine, then atmospheric pressure would drive in the piston as a spray of cold water condensed the steam and created a vacuum. Newcomen's design was improved by the Scottish engineer James Watt in 1765. The invention of the steam engine revolutionized shipping for a relatively brief time in history.

1714 • The British Board of Longitude established a contest in which £20,000 would be awarded to the first person who developed a means for a ship to find its longitude, anywhere on Earth, within an accuracy of 0.5 degree (which is equal to 30 minutes of longitude).

1714 • The German physicist Gabriel D. Fahrenheit invented the first mercury thermometer of the type still in use today.
See also 1593; 1641; 1709.

1715 • The first leather diving suit was introduced. It had a large, round helmet with one window and three hoses, one for outgoing air and two for incoming air, pumped in by bellows from the surface.

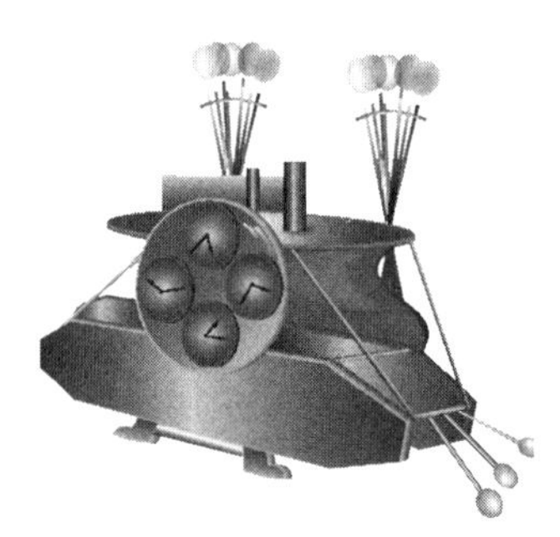

Harrison chronometer

octant (Courtesy of NOAA)

1725 • The Italian naturalist LUIGI MARSIGLI FERDINANDO created a book featuring the whole of marine science as it was known at the time. The book's title was *Histoire physique de la mer.*

1730 • The English mathematician John Hadley and the American inventor Thomas Godfrey invented the octant, which, by measuring the altitude of celestial bodies, could work out a ship's position in latitude but not longitude.

1735 • The English inventor and horologist JOHN HARRISON completed a working prototype of the first accurate marine chronometer. It is not possible to calculate the true position of a ship through latitude and longitude determinations without accurate timekeeping. This invention improved navigation forever.

1747 • The English naval surgeon James Lind introduced the concept of providing fresh fruit aboard vessels to prevent scurvy, a disease that for generations had taken many lives of crew members throughout their long voyages at sea.

1757 • The Scottish inventor John Campbell designed the sextant, enabling seafarers to measure both latitude and longitude to determine a ship's position accurately. It was an improvement on the design of the OCTANT.

1768–1778 • Over a 10-year period, the English navigator CAPTAIN JAMES COOK sailed thousands of miles of coast, conducting surveys and charting his findings. His meticulous mapmaking and pioneering voyages where no European had gone before opened the waters for future exploration. He was also the first ship's captain who, by providing fresh fruits and sauerkraut, prevented scurvy aboard ship.

1772 • The English priest-turned-scientist Joseph Priestley—inventor of soda pop and pencil erasers—conducted studies that led to the documentation of a process that came to be known as PHOTOSYNTHESIS.

1788 • The American inventor John Smeaton made the first modern improvements on the dive bell by implementing cast iron in its construction, a hand pump for easier supply of air to the diver, and an air reservoir system and nonreturn valves to prevent air from being sucked back up the hoses during situations when the pump was inactive.

sextant (Courtesy of NOAA)

1800–1804 • The French naturalist FRANÇOIS PERON circumnavigated the globe, making random and uncertain deep temperature measurements of the sea. The data collected were minimal, but his enthusiasm for oceanography helped to foster interest in marine science.

1803–1806 • The Russian navigator Ivan F. Kruzentein led a Russian circumnavigation of the globe and with the help of J. C. Horner made numerous significant deep ocean temperature measurements in the Sea of Okhotsk and the South Pacific.

1810 • Beginning in this year, the English scientist WILLIAM SCORESBY spent 12 years conducting deep ocean and surface observations of the Arctic waters near Spitsbergen and Greenland.

1815 • The German explorer OTTO VON KOTZEBUE set out on his first circumnavigation of the world, during which many deep-sea observations were recorded.

1817–1818 • The British explorer of the Arctic SIR JOHN ROSS discovered in seabed samples living organisms at a depth of 5,900 feet (1,798 m) near Baffin Island, Canada.

1819–1821 • The Russian explorer FABIAN VON BELLINGSHAUSEN set sail with the goal of circumnavigating the Antarctic continent on the first expedition to accomplish that feat. Though he sorely lacked a naturalist, many valuable observations of animals, land, and sea were recorded.

1823 • The German explorer OTTO VON KOTZEBUE set out on his second circumnavigation of the world, this time with the Russian physicist EMIL VON LENZ. During this expedition it was recognized that surface water flowing from low to high latitudes must originate in deep-water flows from the Poles.

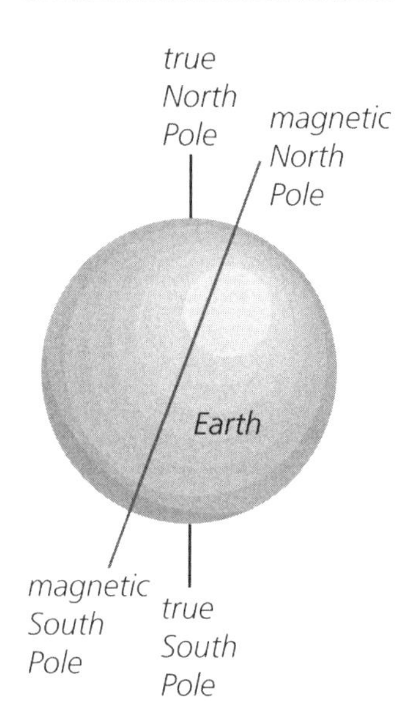

magnetic north

1826 • The French mathematician Charles Sturm and the Swiss physicist Daniel Colladon clocked the speed of sound through water at 4,708 feet (1,435 m) per second, nearly five times faster than the speed sound travels through the air.

1826–1836 • The famous *Beagle* expedition, named for the research vessel the HMS *Beagle,* captained by the British hydrographer and naval officer ROBERT FITZROY, occurred. Between 1831 and 1836 the British scientist and naturalist CHARLES DARWIN sailed on the *Beagle.* During the voyage he correctly worked out the subsidence theory behind the formation of coral atolls, classified barnacles, and developed his famous theory of natural selection.

1828 • The Irish amateur naturalist J. Vaughan Thomson conducted studies of marine plankton off the Irish coast and was the first to describe the planktonic stages of young crab larvae (called zoeae) as they pass through their various free-swimming and molting stages.

1830 • The American inventor Charles Condert created a style of SCUBA apparatus whereby the diver's air supply was stored in copper tubing worn wrapped around the body. The air was to be released inside a hood worn by the diver. Condert died while testing his own invention.

1831 • The British explorer JAMES CLARK ROSS, while on an Arctic expedition with his uncle, Sir John Ross, discovered the magnetic North Pole.

USS *Vincennes* (Courtesy of NOAA)

1838–1842 • The war sloop USS *Vincennes,* commanded by CAPTAIN CHARLES WILKES, was one of four ships assigned by the U.S. Congress to conduct an exploratory expedition to "extend the bounds of science and to promote knowledge," as well as enhance current navigation information and thus sustain commerce.

1839–1843 • The British explorer JAMES CLARK ROSS, nephew of JOHN ROSS, along with the British botanist Joseph D. Hooker, during dredging operations in Antarctic waters discovered abundant life at depths of 23,000 feet (7,010 m). Hooker, at the same time, made another important discovery, that of microscopic plants called diatoms. He proposed with certainty that they played a vital role in the marine food chain. Later, his ideas were found to be correct.

1847 • The British botanist Joseph D. Hooker, close friend of CHARLES DARWIN, noted that planktonic diatoms were plants and proposed that they play an ecological role in the sea similar to that of photosynthetic plants on land.

1850 • An organization called the Royal Meteorological Society of England was founded. Their goal is to advance meteorological science by means of journals and other publications, discussion meetings, conferences, professional accreditation, grants, medals, prizes, workshops for schoolteachers, and other educational activities. The science of meteorology includes climatology, hydrology, physical oceanography, and other related disciplines.

See also Useful websites in the Appendixes.

1854 • The British naturalist SIR EDWARD FORBES published the book *Distribution of Marine Life.*

1855 • The American naval officer and oceanographer MATTHEW FONTAINE MAURY thought of by many as the "father of oceanography," published the first textbook of modern oceanography, *The Physical Geography of the Sea.* This book detailed the study of ocean currents and prevailing winds, and seafarers instantly recognized its usefulness.

1863 • The NAS (National Academy of Sciences) was formed as an organization dedicated to the furtherance of science and technology, providing independent, objective scientific advice to the nation. Their Environment and Earth Sciences Program Unit sponsors oceanographic research.

See also Useful websites in the Appendixes.

1869 • The Suez Canal in Egypt opens, making shipping possible between the Red Sea and the Mediterranean. This cut transit time from India to Europe nearly in half.

1871 • The ROYAL SOCIETY OF LONDON requested the British government to raise funds for investigating the world's oceans, sparking the first era of marine research.

1872–1876 • The era of exploration is considered to have begun when the ship HMS *Challenger,* carrying five scientists under the direction of the Scottish naturalist CHARLES WYVILLE THOMSON undertook a journey on which extensive biological, chemical, geographical, and physical observations were conducted throughout the Atlantic and Pacific Oceans; this voyage, called the Challenger expedition, was one of the most famous research voyages ever dedicated to science. Findings were published in the celebrated *Challenger Reports* and presented the first broad view of the character of the oceans.

1873 • The first American Marine Biological Research Laboratory was established in the John Anderson School of Natural History on Penikese Island, Massachusetts, United States, by the Swiss-American naturalist LOUIS AGASSIZ. The laboratory was moved to Woods Hole, Massachusetts, in 1888.

1873 • The Scottish naturalist CHARLES WYVILLE THOMSON published one of the first textbooks on oceanography, *The Depths of the Sea,* in which he recorded results of his biological and oceanographic studies conducted between 1868 and 1870 during deep-sea dredging operations.

1873 • The German zoologist ANTON DOHRN established the early marine biology station in Naples, Italy, the Stazione Zoologica, which was among the first scientific facilities of its kind to accept visiting scientists from other countries.

1874–1876 • The German ship SMS *Gazelle,* as had the *Challenger,* added greatly to the knowledge of ocean physics by conducting studies in the Atlantic and Pacific Oceans.

1875 • The American zoologist born in Switzerland ALEXANDER AGASSIZ established an aquarium and laboratory at Newport, Rhode Island, which operated until 1910.

1875–1876 • The ship USS *Tuscarora,* commanded by Capt. George E. Belknap, was the first ship to use a piano wire as a sounding line (though WILLIAM THOMSON had used one on his yacht) and obtained a sounding more than five miles (8 km) deep off the coast of Japan in the JAPAN TRENCH. On this voyage as well, studies were made on the distribution of temperature in the northern Pacific Ocean.

1876 • The English inventor Henry Fleuss developed a compact SCUBA system that employed pure compressed oxygen in its operation and a system to recirculate the exhaled air, again and again, after carbon dioxide had been removed. Breathing pure oxygen at depth, however, poses an almost certain danger that the oxygen will become toxic.

1877 • The English mathematician John William Strutt, better known as Lord Rayleigh, published his *Theory of Sound,* pioneering the study of modern acoustics.

1877–1880 • Scientists aboard the ship USS *Blake* explored marine life off the coast of Florida, the Gulf of Mexico, and the Caribbean Sea under the direction of ALEXANDER AGASSIZ and later John Elliot Pillsbury (from 1885). On the voyage a remarkable series of current measurements were made in the Florida Straits and of passages within the Windward Islands. It was aboard the *Blake* that Agassiz developed the Agassiz trawl, which consisted of a cylindrical net with an iron hoop at each end, all suspended from an iron frame. It was a great improvement on sampling nets of the past.

1878 • The long-sought NORTHEAST PASSAGE was opened by NILS A. E. NORDENSKJÖLD when he sailed from the Atlantic Ocean to the Pacific Ocean by way of the Arctic Ocean through a route along the northern coast of Europe and Asia.

1884 • England established the Marine Biological Association to study the ecology of the coast and to promote public understanding of marine biology in its broadest sense.

1885 • Scientists aboard the USS *Blake,* commanded by John Elliot Pillsbury, made remarkable current and temperature measurements at different depths across the Florida Strait, waters that separate the southern tip of Florida and the island of Cuba, and at a number of sites throughout the passages of the Windward Islands, a chain forming part of the West Indies at the southern end of the Lesser Antilles.

1885–1926 • The Royal Indian Marine survey ships HMS *Investigator* (1885–1905) and *Investigator II* (1908–26), conducted extensive, mainly biological, studies of the Bay of Bengal and the Arabian Sea. Broad collections and reports were made on undersea organisms, such as the alcyonarian corals, or SOFT CORALS, adding greatly to knowledge of aquatic life in these regions.

1886–1889 • The Russian research vessel *Vitaz,* under the command of Adm. STEPAN O. MAKAROV, circumnavigated the globe. Markov made many observations of sea temperature, specific gravity, currents, and tides.

1886–1922 • In these years the Monegasque oceanographer PRINCE ALBERT I OF MONACO commissioned the vessels *Hirondelle* (1885–88), *Princess Alice* (1892–97), *Princess Alice II* (1898–1907), and *Hirondelle II* (1911–22) to conduct systematic oceanographic studies from the Cape Verde Islands to Spitsbergen and from the Mediterranean to the coasts of New England and Newfoundland, during which many significant deep-sea collections were made.

1887 • The USS *Albatross* was first commissioned for oceanographical work by the U.S. Fish and Fisheries Commission and undertook voyages through 1925. While working in the tropical Pacific the crew took aboard more deep-sea fishes in one haul of the dredge than did the HMS *Challenger* during its entire cruise.

***Albatross* (Courtesy of NOAA)**

1888 • The Marine Biological Association established a marine biological laboratory in Plymouth, England.
See also 1884.

1888 • The Marine Biological Laboratory was established at Woods Hole, Massachusetts, with just two instructors and eight students.
See also 1873.

1889 • The year of the German Plankton Expedition, conducted aboard the SS *National* by Professor VICTOR HENSEN, whose objective was to study the plankton and food cycles of the ocean.

1892–1896 • The Norwegian ship *Fram,* specially constructed to withstand potential entrapment in pack ice, under the direction of FRIDTJOF NANSEN, drifted across the Arctic Basin, making valuable observations and confirming that there was no Arctic continent. Nansen made many oceanographic, magnetic, astronomical, and meteorological observations and showed that large volumes of comparatively warm, highly saline surface waters were carried to the Arctic from lower latitudes.

Fram **(Archival Photograph by Mr. Steve Nicklas/Courtesy of NOAA)**

1895–1896 • The Danish survey ship *Ingolf* cruised off Iceland in the North Atlantic, making extensive biological and physical studies.

1895–1898 • In his yacht the *Spray,* the American Joshua Slocum became the first to circumnavigate the world alone.

1896 • Generated by an undersea earthquake, a tsunami wave 125 feet (38 m) high laid to waste 175 miles (281 km) of the Japanese coast, killing and wounding more than 30,000 people. In terms of life and limb, it was the sea's most devastating act in that century, in that region, proving yet again the mastery of the oceans over the planet.

1897–1899 • The Belgian ship *Belgica,* while on expedition conducting physical observations and making biological collections around Peter I Island, became the first ship to winter in pack ice in the Antarctic Ocean. The voyage was led by the Belgian explorer Adrien de Gerlache de Gomery. Among the crew was the Norwegian explorer ROALD AMUNDSEN.

1898–1899 • The German survey ship *Valdivia,* commanded by Professor Carl Chun, conducted a deep-sea expedition into the high southern latitudes of the Indian, Atlantic, and

Antarctic Oceans, trawling successfully at great depths, retrieving many deep-sea collections, and studying the vertical distribution of pelagic organisms.

1899–1900 • The Dutch survey ship *Siboga* conducted the Netherlands Deep Sea Expedition, led by Professor Max Weber, to study the hydrologic and biologic features of the waters of the Dutch East Indies and Malaya. The expedition gathered an enormous collection of both shallow and deep-water organisms; the data were later compiled in many volumes.

1901–1903 • Germany launched their first Antarctic expedition, led by Erich von Drygalski aboard the *Gauss,* during which the Antarctic Convergence was first discovered. They reached as far as 66° S in the western Indian Ocean and conducted extensive studies of all aspects of oceanography.
See also ANTARCTIC CONVERGENCE.

1902 • The Danes established the ICES (International Council for the Exploration of the Sea), an international science organization that comprised 19 member countries to study and therefore help to safeguard the marine ecosystems of the North Atlantic Ocean. Today it is the oldest intergovernmental marine science organization in the world and a leading forum for the promotion, coordination, and distribution of research on the physical, chemical, and biological systems in the North Atlantic Ocean and advice on human impact on its environment, in particular fisheries effects in the Northeast Atlantic.
See also Useful websites in the Appendixes; ICES.

1902 • The Scottish Antarctic Expedition, led by Dr. William S. Bruce, sailed into the Southern Ocean in the refitted whaling vessel *Scotia,* their primary directive to conduct extensive hydrographical work in the WEDDELL SEA and survey the South Orkney Islands and study the wildlife there. The expedition enjoyed success in dredging at great depths in the Weddell Sea and off Coats Land, named for the expedition's first sponsors, and made extensive scientific studies on penguins.

1903 • America established the independent biological research laboratory, Scripps (Scripps Institution of Oceanography) in La Jolla, California, which in 1912 became part of the University of California, San Diego. In 1925 it was named after the Scripps family in recognition of their generous financial support. Its research programs include biological, chemical, physical, geological, and geophysical studies of the oceans. Doctoral degrees are offered in many of these areas of learning.

1903 • The Monegasque oceanographer PRINCE ALBERT I OF MONACO, a key participant, established a chart series illustrating measurements and charting of the ocean floor. This series was called GEBCO (General Bathymetric Chart of the Oceans). Today, GEBCO charts are available on CD-ROM; ordering information can be found at their website.

See also Useful websites in the Appendixes; GEBCO.

1903–1906 • The long sought NORTHWEST PASSAGE was opened by the Norwegian explorer ROALD AMUNDSEN, when he sailed from the Pacific Ocean to the Atlantic Ocean by way of the Arctic Ocean through a route along the northern Canadian archipelago and Alaska.

1904–1913 • Norway sponsored the *Michael Sars* Expedition, which conducted many samplings of mesopelagic and bathypelagic organisms from the North Atlantic.

1905–1929 • The ships *Galilee* (1905–09) and *Carnegie* (1909–29) of the Carnegie Institute of Washington made the most extensive studies of the Earth's magnetic field over the ocean that had ever been achieved. In the latter years of the voyage, extensive oceanographic observations were concentrated in the Central Pacific Ocean.

1906 • The Italian airship designer Enrico Forlanini built the first successful hydrofoil, a fast craft that rises above the water on blades. Today, hydrofoils are commonly seen around the world in service as ferries and passenger craft.

1906 • PRINCE ALBERT I OF MONACO established the Oceanographic Museum and Aquarium in Monaco to house and display his collections from his research expeditions. For 30 years, from 1957, JACQUES COUSTEAU headed the museum.

1906 • A German expedition aboard the SMS *Planet* conducted surveys in the Indian and Atlantic Oceans, during which the deep Java Trench was discovered. It was recorded as being over 23,000 feet (7,010 m) in depth.

1908–1910 • The Danish biologist JOHANNES SCHMIDT carried out the first major oceanographic expedition of the new century.

1911–1912 • Germany launched its second south polar expedition in an attempt to make a trans-Antarctic crossing. The expedition's leader was the German explorer Wilhelm Filchner. The ship was the *Deutschland,* its captain Richard Vahsel. Filchner commissioned a team of scientists including an astronomer, two doctors, and the oceanographer W. Brenneck. Though the crossing was not a success, the expedition added greatly to physical knowledge of the South Atlantic and the WEDDELL SEA and to knowledge of plankton.

1912 • The German geologist and meteorologist ALFRED L. WEGENER suggested the idea of CONTINENTAL DRIFT, which at first was ridiculed. In 1915 he detailed the theory in *Origins of Continents and Oceans.*

1919 • The German scientist Dr. Hannes Lichte advanced the understanding of underwater acoustics through his theory that sound waves moving through water refract either downward or upward when they encounter changes in temperature, salinity, and pressure.

1920 • The English botanist Agnes Arber (1879–1960) published a major book of the times, *Water Plants: A Study of Aquatic Angiosperms,* which became the standard compilation of information on aquatic botanicals.

1921 ● Denmark sponsored their famous *Dana I* and *Dana II* Expeditions, led by the Danish biologist JOHANNES SCHMIDT, during which many collections were made at great depths at many areas around the globe. The research took place over a span of 15 years.

1925–1927 ● The German oceanographer GEORG WÜST, aboard the RV *Meteor,* directed oceanographic studies in the South Atlantic Ocean, employing for the first time many research techniques used today. At that time their work contributed more to the understanding of physical, chemical, and geological understanding of the ocean than any previous expedition.

1925–1939 ● The British carried out extensive research around the Southern Ocean, which they called the *Discovery* Investigations. The results were published as the *Discovery Reports* and established modern marine research in the Antarctic region. The ship *Discovery* (1926–31) set sail to survey the whaling grounds around South Georgia Island and assess the status of the stock. The chief zoologist for the voyage was the English marine biologist Sir Alister Hardy. Alone and then with the sister ship, *William Scoresby* (1926–39), soundings, hydrographic surveys, observations, and quantitative plankton studies were made. In 1929 a third ship, *Discovery II,* was introduced to the expedition. Deep dredging drew to the surface a wide variety of organisms including bioluminescent krill and species of octopus never seen before. The mountain of data gathered was reported and published.

1927 ● The British biologist and naturalist Charles S. Elton (1900–91) published two landmark books: *Animal Ecology* (1927) and *Animal Ecology and Evolution* (1930). His influence is far-reaching and ecologically is still felt today, as his work on the succession of organisms gave rise to the first model of the food chain.

1927–1931 ● The Norwegian Antarctic Expedition took place aboard the *Norwegia* under the direction of Professor Haakon

Mosby and Gunner Isachsen, whose extensive observations greatly expanded knowledge of the oceanographical and biological intricacies of the Southern Ocean.

1928 • The English marine biologist Sir Alister Hardy, chief zoologist for the 1926 *Discovery* Expedition, founded a Zoological Department at the University of Hull, England.

1928–1930 • The Danish sponsored the Carlsberg Foundation Oceanographical Expedition, a circumnavigation, which took place aboard the *Dana* under the direction of JOHANNES SCHMIDT. It was the largest of many voyages led by Schmidt.

1929–1931 • Inspired by the German Atlantic Expedition of the RV *Meteor,* the Dutch *Willebrord Snellius* Expedition took place in the Southwest Pacific and East Indian Oceans and conducted studies on biological samplings, such as those of FORAMINIFERA. The expedition also carried out highly detailed oceanographic surveys of the waters surrounding the East Indian Archipelago.

1930 • WHOI (Woods Hole Oceanographic Institution) was established by the National Academy of Sciences at Cape Cod, Massachusetts, supported by a $3 million grant from the Rockefeller Foundation. It was in this year that its research vessel *Atlantis* was first put into service.

1931 • A nonprofit organization, ICSU (International Council of Scientific Unions), was founded to assemble groups of natural scientists in international scientific ventures. A scientific research council of 98 multidisciplinary national members and 26 international single-discipline unions coordinate programs that address major international, interdisciplinary scientific issues that would be very difficult to manage single-handedly.

See also Useful websites in the Appendixes.

1931 • The British-built Italian vessel *Conte di Savioa,* a passenger liner, became the first ship to be fitted with gyroscopic stabilizers. These were movable fins attached below the waterline to prevent ships from rolling while under way.

1933 • The John Murray Expedition, aboard the Egyptian research vessel *Mabahiss,* set sail for the Indian Ocean, where they took biological and geological samplings, conducted extensive soundings, and made physical and chemical analyses. Present on the expedition was the English scientist William Ian Farquharson, who drafted a report on bathymetry.

1934 • The American explorer, naturalist, and zoologist WILLIAM BEEBE, along with OTIS BARTON, descended to a depth of 3,027 feet (923 m) in a tethered bathysphere to observe marine life.

1937 • The British engineer Athelstan Spilhaus designed and built the BATHYTHERMOGRAPH, a device used to measure the temperature of seawater at various depths.

1941–1942 • The largest battleships ever designed were completed. The sister ships *Yamato,* built at Kure, Japan, and *Musashi,* built at Nagasaki, Japan, were 865 feet (263 m) long and 121 feet (37 m) wide; displaced 71,659 tons (72,805 mtn), loaded; and had a top speed of 27 knots (29 mph [46 kmh]). They were both sunk by the U.S. Navy carrier aircraft fire in 1944 *(Musashi)* and 1945 *(Yamato).*

bathysphere

1942 • The classic marine science book *The Oceans* was published as the first modern textbook in oceanography, a collaborative work by H. SVERDRUP, R. Flemming, and M. Johnson.

1943 • The French oceanographer JACQUES COUSTEAU, along with the French engineer Émile Gagnan, developed for use in diving the automatic demand valve system, the AQUA-LUNG, more commonly known as SCUBA gear.

1945 • In collaboration with 37 countries, UNESCO (United Nations Educational, Scientific and Cultural Organization) was established in London. Its initial goal was to establish peace and security among nations through their combined education, science, and culture. This organization is responsible, among many other things, for the International

Indian Ocean Expedition that began in 1959 and for research on coastal zone management. Their headquarters is located in Paris, France, and 73 field offices can be found around the world.

See also Useful websites in the Appendixes.

1946 • Under the support of the United Nations, the IWC (International Whaling Commission) was founded with the objective of regulating coastal whaling industries and the associated pelagic whaling fleets.

See also Useful websites in the Appendixes.

1947 • In an attempt to prove that Polynesians could have migrated from South America, THOR HEYERDAHL sailed across the Pacific Ocean in a balsa raft named the *Kon Tiki.* He successfully navigated from Peru to the Tuamotu Islands.

1949 • The U.S. Government inaugurated three major programs in ocean policy: the Gulf States Marine Fisheries Commission, whose principal objectives are the conservation, development, and full utilization of the fishery resources of the Gulf of Mexico; the International Convention for the Northwest Atlantic Fisheries Commission, whose principal objectives are the investigation, protection, and conservation of the fisheries of the Northwest Atlantic Ocean; and the Inter-American Tropical Tuna Commission, responsible for the conservation and management of fisheries for tunas and other species taken by fishing vessels in the eastern Pacific.

"Prince Axel's wonder-fish"

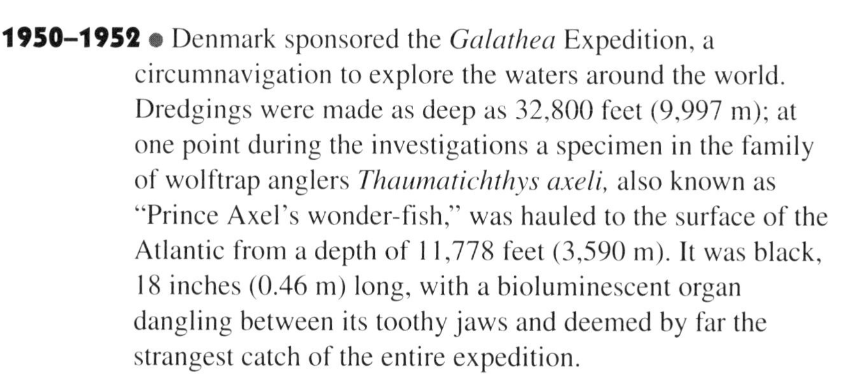

1950–1952 • Denmark sponsored the *Galathea* Expedition, a circumnavigation to explore the waters around the world. Dredgings were made as deep as 32,800 feet (9,997 m); at one point during the investigations a specimen in the family of wolftrap anglers *Thaumatichthys axeli,* also known as "Prince Axel's wonder-fish," was hauled to the surface of the Atlantic from a depth of 11,778 feet (3,590 m). It was black, 18 inches (0.46 m) long, with a bioluminescent organ dangling between its toothy jaws and deemed by far the strangest catch of the entire expedition.

1958 • The first American nuclear-powered submarine, USS *Nautilus,* developed by the U.S. Navy admiral Hyman G. Rickover, became the first submarine to reach the North Pole by traveling *under* the sea ice.

1959–1965 • The IOC (Intergovernmental Oceanographic Commission) of UNESCO (United Nations Educational, Scientific, and Cultural Organization) coordinated a multination, multiship campaign, IIOE (International Indian Ocean Expedition), whose goal was to collect data in the Indian Ocean. The findings of their combined efforts modernized scientific knowledge of those waters.
See also Useful websites in the Appendixes.

1960 • The Swiss scientist AUGUSTE PICCARD's bathyscaphe *Trieste,* piloted by his son, JACQUES PICCARD, and Donald Walsh, descended to the deepest part of Earth's oceans, reaching a recorded depth of 35,810 feet (10,915 m) in the Pacific's MARIANA TRENCH off Guam.

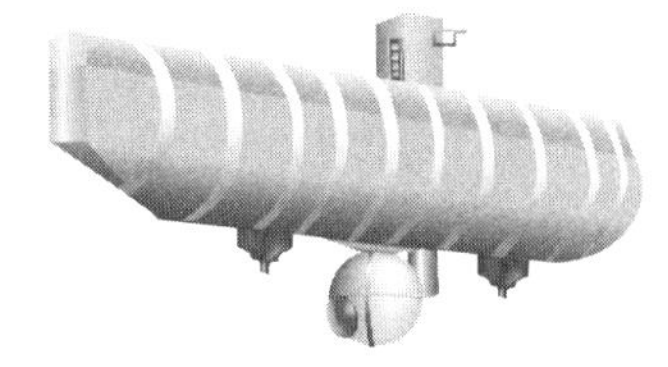

Trieste

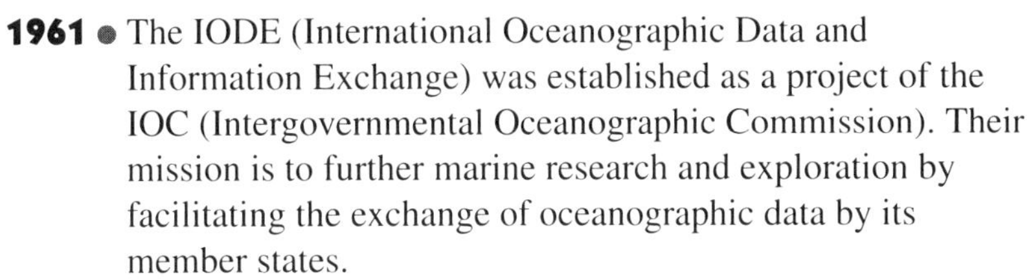

1961 • The IODE (International Oceanographic Data and Information Exchange) was established as a project of the IOC (Intergovernmental Oceanographic Commission). Their mission is to further marine research and exploration by facilitating the exchange of oceanographic data by its member states.
See also Useful websites in the Appendixes.

1961–1966 • Project Mohole, led by HARRY H. HESS, became the first expedition to drill through the Earth's oceanic crust to the mantle beneath, as far as 601 feet (183 m) into the seabed beneath 11,700 feet (3,566 m) of water. Mohole's goal was to retrieve a sample of material from the Earth's mantle by drilling into the boundary between the Earth's crust and upper mantle, called the Mohorovicic discontinuity, or Moho.

1962 • A marine mammal facility sponsored by the U.S. Navy was established at Point Mugu, California.

1963 • A multidisciplinary ocean and marine network, the Marine Technology Society of Columbia, Maryland, was incorporated with the purpose of giving members in academia, government, and industry a common forum for the exchange of information and ideas and to "make real contributions in fostering progress in the marine sciences for the benefit of mankind."

See also Useful websites in the Appendixes; MTS.

1964 • The deep-sea submersible *ALVIN*, of WHOI (Woods Hole Oceanographic Institution), was commissioned. During the years since, *Alvin* has conducted over 1,000 research dives, launched mainly from the tender *Atlantis II,* which was retired from service in 1996.

1965 • In a program of the U.S. Navy called Sealab II, a dolphin worked in the open ocean off La Jolla, California, carrying tools and equipment from the sea surface to divers working 200 feet (61 m) below, thus demonstrating that marine mammals could do useful work, untethered, in open waters.

1965–1966 • NASA (National Aeronautical and Space Administration) launched the Gemini space program. Throughout its 10 crewed missions, extensive collections of photographic images of the Earth were taken. These photos led to recognition the value of space photography in the fields of oceanography, geology, and meteorology.

Hydrolab (Courtesy of NOAA/NURP)

1966 • The underwater research habitat Hydrolab was constructed; beginning in 1970, it served as a shallow water underwater laboratory for science missions. Located in Salt River Canyon, Saint Croix, United States Virgin Islands, this facility was funded by the Perry Foundation and the precursor to the NOAA (National Oceanic and Atmospheric Administration). It was a sparse, cylindrical chamber 16 feet (4.8 m) long and eight feet (2.4 m) wide, which served as a base for hundreds of aquanaut researchers from nine different countries. Its scientific missions produced a 90 percent publication per mission productivity level. The habitat was decommissioned in 1985 and placed on display at the Smithsonian Institution's National History Museum in Washington, D.C. The Aquarius, a more flexible and technologically advanced habitat system, has replaced the Hydrolab as NOAA's principal seafloor research laboratory.

1966 • The U.S. Navy, using the CURV (cable-controlled underwater recovery vehicle), recovered an atomic bomb lost off Palomares, Spain, after an aircraft accident, generating attention to ROVs (remotely operated vehicles).

1968–1975 • The United States National Science Foundation launched a deep-sea drilling project and confirmed the theory of seafloor spreading.

1970 • NOAA (National Oceanic and Atmospheric Administration), an agency of the U.S. Department of Commerce, was created. Included in this organization are the National Ocean Service; the National Weather Service; the National Marine Fisheries Service; the National Environmental Satellite, Data, and Information Service; and the Office of Oceanic and Atmospheric Research.

1972 • UNOLS (University-National Oceanographic Laboratory Systems) was established as a consortium of 57 academic institutions and national laboratories that participate in significant marine science programs that either operate or use the U.S. academic research fleet. The 27 UNOLS research vessels constitute the largest and most capable fleet of oceanographic research vessels in the world. It is now an organization of 64 institutions and laboratories that coordinate oceanographic ships' schedules and research facilities. Their offices are located at Moss Landing, California.

See also Useful websites in the Appendixes.

1972 • The U.S. Congress passed the Marine Protection, Research and Sanctuaries Act; the Clean Water Act; the Marine Mammals Protection Act; and the COASTAL ZONE MANAGEMENT Act, as the negative impacts of human activity, such as oil spills and unregulated harvesting of marine life, were recognized.

1973 • An international treaty called CITES (Convention on International Trade in Endangered Species of Wild Fauna and Flora) was drawn up to protect wildlife from overexploitation.

1973 • With only a few minutes of air remaining, the United States Navy, using the CURV (cable-controlled underwater recovery vehicle), rescued the pilots of the sunken submersible *Pisces* off the coast of Ireland.

1974 • The Argentine Oceanographic Data Center (CEADO [Centro Argentino de Datos Oceanograficos]) was established by an agreement between the SHN (Naval Hydrographic Service) and CONICET (National Scientific and Technical Research Council), of Argentina, to support marine science research and development by providing oceanographic data and information to the national and international scientific community and other users who request information.

See also Useful websites in the Appendixes.

1974 • The Bigelow Laboratory for Ocean Sciences was established in West Boothbay Harbor, Maine. Their mission is to research the biological characteristics of the ocean, including detailed analysis of zooplankton and phytoplankton.

See also Useful websites in the Appendixes.

1974 • The Scottish inventor Stephan Salter was among the first to develop a wave-powered generator as a way to produce power from a renewable resource.

See also WAVE POWER.

1976 • The Magnuson Fishery Conservation and Management Act was enacted, to extend federal fishery jurisdiction to 200 miles (322 km) offshore and establish the current federal fishery management regime.

See also 1992.

1976 • The deepest dive performed in an armored one-atmosphere JIM SUIT occurred off the coast off Spain to recover a television cable; the diver descended to 1,444 feet (440 m).

1977 • SCRIPPS INSTITUTION OF OCEANOGRAPHY launched an uncrewed research sled called *ANGUS* (acoustically navigated underwater survey system) and made the first detailed observations of deep-sea vent communities.

1978 • *Seasat-A,* the first remote-sensing oceanographic satellite, was launched by Jet Propulsion Laboratory, California, with the objective to study Earth's oceans by using SAR (synthetic aperture radar).

1980 • JISAO (Joint Institute for the Study of Atmosphere and the Ocean) was formed as a union between the University of Washington, Seattle, Washington, and NOAA (National Oceanic and Atmospheric Administration) as a research project covering climate variability, global environmental chemical characteristics, estuaries, and recruitment of fish stock in the Pacific Ocean. Key members of JISAO are the Department of Atmospheric Sciences (University of Washington), the School of Oceanography (University of Washington), the Pacific Marine Environmental Laboratory (NOAA), the National Weather Service (NOAA), and the National Marine Fisheries Service (NOAA).

See also Useful websites in the Appendixes.

1982 • The U.S. Government enacted CBRA (Coastal Barriers Resources Act), ocean policy that restricts any financial support, federal or otherwise, of development of coastal barriers that might lead to damage of property, fish, wildlife, and other natural resources.

1985 • The American deep-sea explorer ROBERT BALLARD, using a camera aboard a robot submersible, located the famous lost wreck of the RMS *Titanic,* which sank on April 14, 1912, about 95 miles (153 km) south of Newfoundland, Canada, on its way to New York.

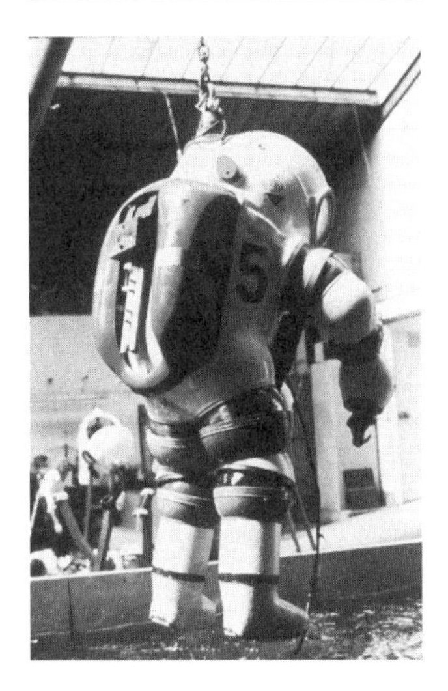

Jim suit (Courtesy of NOAA/NURP)

ANGUS (Photograph by quartermaster Joseph Schebal/ Courtesy of NOAA)

1986–1988 ● NOAA (National Oceanic and Atmospheric Administration) launched the Beaufort Sea Mesoscale Project, whose mission was to investigate oceanic circulation over the Beaufort Sea Shelf and its interaction with meteorological phenomena.

1987 ● The Center for Sea and Atmosphere Research was established at the University of Buenos Aires, Argentina. Also known as CIMA (Centro de Investigaciones del Mar y la Atmósfera), it has the goal of broadening knowledge of the physical processes that control and determine the behavior of the ocean and the atmosphere.

See also Useful websites in the Appendixes.

1988 ● A nonprofit organization, the Oceanography Society, was founded in Washington, D.C. Its purpose is to circulate oceanographic knowledge and its application through research and education. It promotes communication among oceanographers and provides a community for the development of agreement throughout the disciplines of the field.

See also Useful websites in the Appendixes.

1988 ● The Center for Hydro-Optics and Remote Sensing was established at San Diego University, California. Its function is to carry out interdisciplinary research on biooptical properties and remote sensing of the ocean environment. It maintains equipment inventories and provides laboratory facilities for shipboard measurements of biooptical variables.

1988–1989 ● The United States, Denmark, Canada, Norway, and France joined to form what was known as CEAREX (Coordinated Eastern Arctic Experiment). Studies were carried out in the Greenland and Norwegian Seas between September 1988 and May 1989, during which data were gathered on wind profiles, humidity, sea ice, acoustic measurements, hydrographic, features, biophysical attributes, and bathymetry.

See also Useful websites in the Appendixes.

1989 • In Alaska's Prince William Sound the tanker *Exxon Valdez* caused the largest oil spill in U.S. history: 11 million gallons (41,635,000 l) of crude oil blanketed 1,200 miles (1,931 km) of coastline, killing an estimated 100,000 birds, including some 150 bald eagles and around 1,000 sea otters, to name a few. Millions of dollars and work hours, involving some 11,000 workers, were spent on cleanup. The damaging impact it had on the environment is an ongoing investigation. This was one of many major oils spills that occurred in the last three decades.

1990 • The international nonprofit association AMRS (Alliance for Marine Remote Sensing) was founded. Their goal is to promote and develop marine applications of remote sensing technologies. AMRS operates the Center for Marine Remote Sensing, located in Bedford, Nova Scotia.

See also Useful websites in the Appendixes.

1992 • The research program PFRP (Pelagic Fisheries Research Program) was established at the University of Hawaii at Manoa, Honolulu, Hawaii, to provide scientific information on pelagic fisheries to WPRFMC (Western Pacific Regional Fishery Management Council). This program revised the Magnuson Fishery Conservation and Management Act of 1976 to include migratory fish.

See also 1976; Useful websites in the Appendixes; PFRP.

1993–1994 • The United States and Indonesia formed a joint oceanographic research project called the Arlindo Project; *Arlindo* is an abbreviation of *Arus Lintas Indonesia,* meaning "throughflow." The primary goals of this project are to study the circulation and water mass stratification of the Indonesian Seas and determine mixing cycles created by meteorological extremes. Studies were conducted from the research vessel *Baruna Jaya I* during the southeast monsoon of 1993 and the northwest monsoon of 1994.

1995 • NOAA (National Oceanic and Atmospheric Administration) conducted COPE (Coastal Ocean Probing Experiment) to determine how environmental conditions affect observations

of the air-sea interface in coastal regions, to improve the related technology, and to evaluate new scattering theories.
See also Useful websites in the Appendixes.

1995 • The New York Zoological Society, during the year of its 100th anniversary, changed its name to the Wildlife Conservation Society. Their purpose, then and now, is to educate the public about matters of zoology and related subjects and the preservation of wildlife.
See also Useful websites in the Appendixes.

1996 • NASA (National Aeronautics and Space Administration's) JPL (Jet Propulsion Laboratory) launched the Advanced Earth Observing Satellite carrying an instrument called a scatterometer. Throughout its 10-month life span, this device recorded more than 190,000 wind measurements per day and continuously mapped the Earth's ice-free oceans every two days, night or day, regardless of weather. The scatterometer provided more than 100 times the ocean wind information ever obtained from ship's reports.

1997 • A program called SAMBA (Sub-Antarctic Motions in the Brazil Basin) was established through the efforts of the French scientists M. Ollitrault and A. Colin de Verdierè. Their initiative was to launch numerous Marvor (Breton for sea horse), floats that would monitor the motion of AAIW (Antarctic Intermediate Water) in the equatorial band and determine how this water mass crosses the equator. A Marvor float can cycle several times between the surface and its predetermined depth. On surfacing, it sends data into space, where a satellite retrieves it and relays it to land-based stations.

1998 • The United Nations declared YOTO (International Year of the Ocean) to acknowledge the importance of the marine environment and to raise public awareness of ocean-related issues.

1999 • NOAA's (National Oceanic and Atmospheric Administration's) deep-sea submersible *ALVIN* explored the unknown depths of the North Pacific Ocean off Alaska, contributing to knowledge of biological communities, which affect the commercial fishing industry. In one area, a known species of crab was discovered that was thought not to exist there, leading scientists to think that the crab fishery is not actually declining but simply moving out of known areas.

2000 • A decision by the IHO (International Hydrographic Organization) delimited a fifth world ocean, called the Antarctic Ocean or Southern Ocean. The ocean extends from the coast of Antarctica north to 60° S latitude, ending along the southern portions of the Atlantic Ocean, Indian Ocean, and Pacific Oceans.

2001 • NASA (National Aeronautics and Space Administration)–French ocean observation satellite *Jason 1,* was launched in December to monitor the circulation of the WORLD OCEAN, study its interaction with the atmosphere, and improve predictions of climactic phenomena.

2002 • A species of lionfish, *Pterios volitans,* was identified as living in Atlantic coastal waters off North Carolina. The existence of this fish, a nonnative of these waters, is under study to determine whether they are reproducing and are able to survive the winter temperatures of these waters. Their beautiful but venomous spines can be deadly to humans. The lionfish range typically within Indo-Pacific waters.

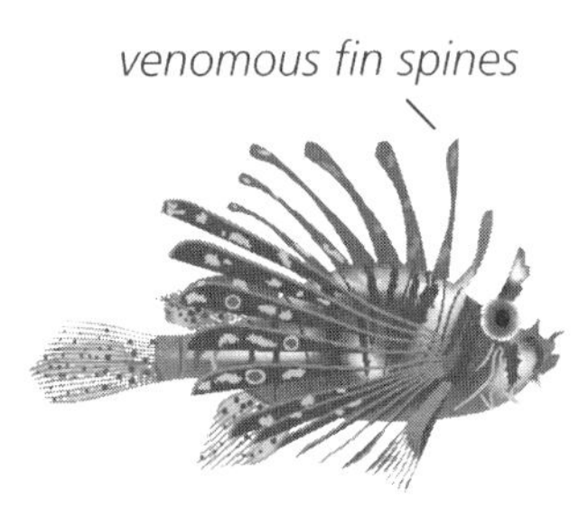

lionfish

2002 • A deadly RED TIDE occurred during February in Elands Bay, off South Africa's western cape. The algal bloom caused a mass stranding of approximately 1,000 tons (907 mtn) of lobsters as, suffocating, they crowded into shallow littoral zones, searching the breakers for oxygen. Consequently they became stranded on the beach as the tide ebbed. The detrimental impact on ocean harvest economy was expected to be felt over subsequent years.

2002 • Weakening of equatorial easterly winds across the Pacific during early spring gave rise to what is known as the EL NIÑO effect, which caused higher than normal sea surface

El Niño

El Niño effect for June 2002

Extreme Drought
Severe Drought
Moderate Drought
Unusually Wet
Very Wet Conditions
Extremely Wet Conditions
Normal Conditions

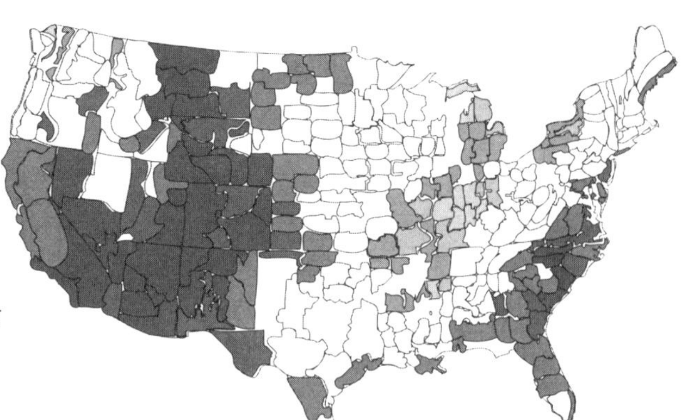

NIÑO effect, which caused higher than normal sea surface temperatures and caused the THERMOCLINE east of the INTERNATIONAL DATE LINE to deepen. These anomalies caused record and near-record weather patterns across the United States, from severe drought in the central, southwestern, and East Coast states to extreme storm and rain conditions extending southward from the Great Lakes region. This phenomenon, which begins at sea, can and does have a huge impact on crop industries, not only in the United States but around the world, leading to severe losses and financial hardship in farming communities.

2003 • September 26 marks the 100th anniversary of Scripps (Scripps Institution of Oceanography) and in this year they dedicate their centennial celebration to conveying to the global population the importance of ocean science.

SECTION FOUR

CHARTS & TABLES

Voyage of Ferdinand Magellan

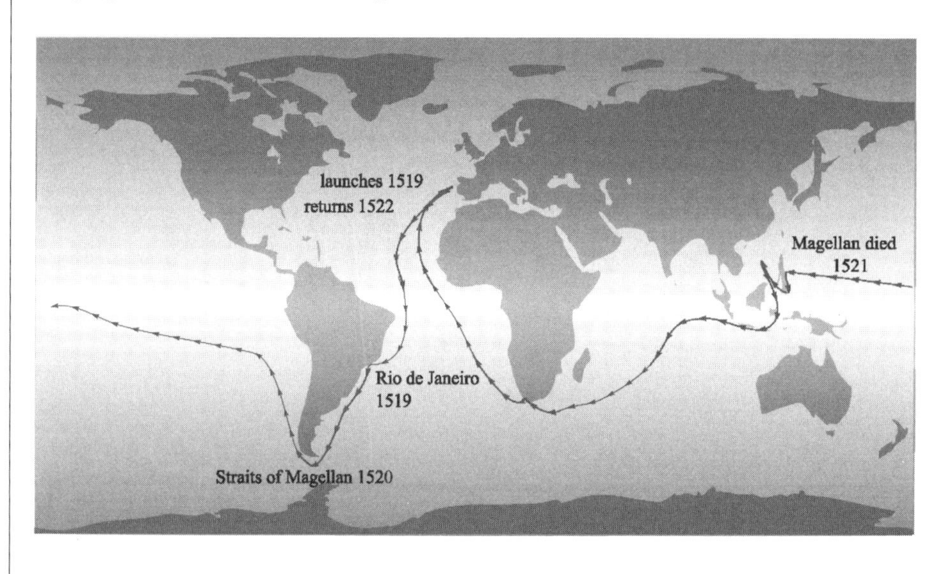

In the late 1500s what is known as the Age of Discovery began. It was during this time that the systematic study of the oceans took hold, beginning with the voyage of Christopher Columbus in 1492.

In 1519, Ferdinand Magellan, in command of the ship *Victoria*, set sail from Spain for a circumnavigation of the globe in order to prove that the world was round.

Of the 241 crew members and five ships that initially set sail, only the *Victoria*, carrying 17 survivors, returned. Magellan, who had died in a battle with island natives in 1521, was not among them.

Voyage of Charles Darwin and Robert Fitzroy

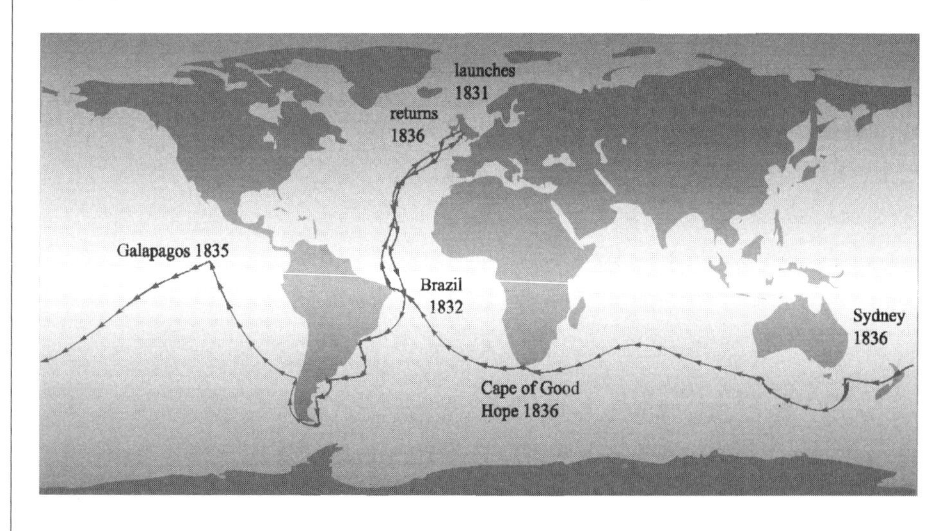

In 1831 a young scientist named Charles Darwin was taken aboard Captain Robert Fitzroy's marine biological and oceanographic research vessel the HMS *Beagle* as an unpaid naturalist.

Throughout the circumnavigation, Darwin and Fitzroy conducted extensive biological and hydrological studies, which resulted in the gathering of detailed information on the oceans, especially coral reefs.

Studies provided Darwin with crucial data for the development of his theory of natural selection.

Fitzroy later collaborated with Darwin on an account of their voyage (1839), *The Voyage of the Beagle*.

Voyage of the *Challenger*

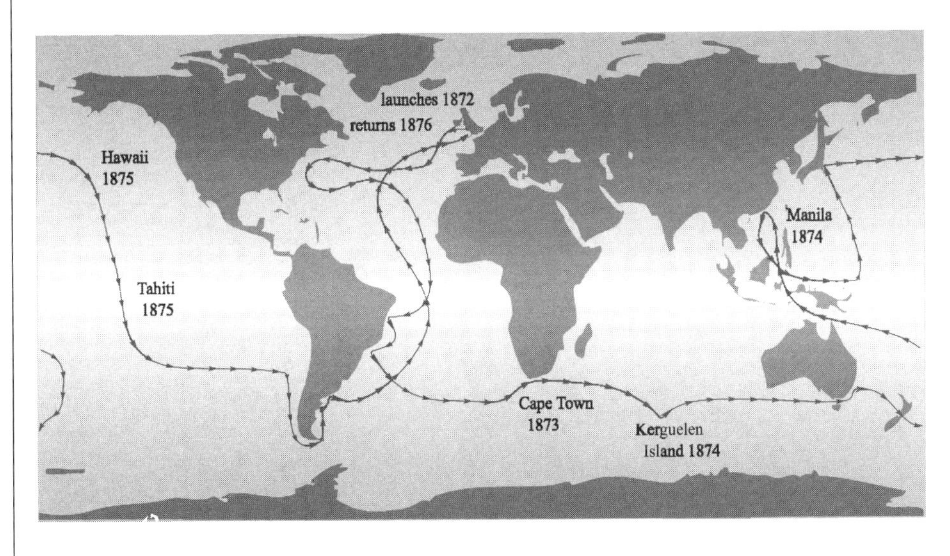

The voyage of the HMS *Challenger* was famous as a cost-effective mission that produced a vast amount of oceanographic data.

With a crew of 270, the Scottish marine biologist Sir Charles W. Thomson set out on a 3.5-year voyage that covered over 68,000 nautical miles. Throughout the voyage, temperature and current readings, dredgings, trawlings, and samplings of marine organisms took place in all three of Earth's major oceans. On the mission's completion, scientists compiled data for over 20 years, becoming what is known as *The Challenger Reports*.

Findings proved beyond doubt that life existed in the ocean at all depths, spurring the launch of many more successful scientific missions.

Drift of the *Fram*

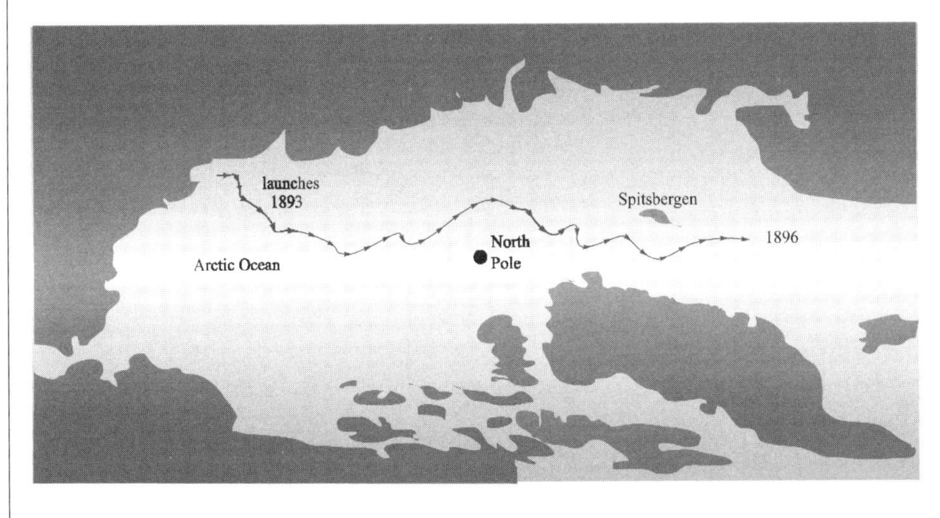

The Norwegian oceanographer Fridjof Nansen radically proposed that only by working with the forces of nature could the North Pole be reached. He claimed that this could be done by allowing one's ship to become trapped in the pack ice and thus drift to the North Pole with the current.

He was not the first to suggest this idea; however, he was the first to attempt it.

Nansen commissioned the building of the sturdy ice ship *Fram*, and in 1893 he set out into the Arctic. He did not reach the North Pole; however, he proved the validity of his idea by drifting icebound farther north than any previous expedition.

World Aquaculture and Commercial Fish Catches, 1995–1999

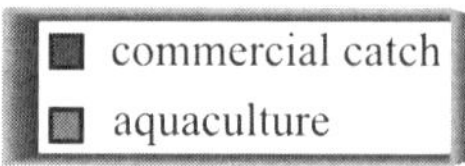

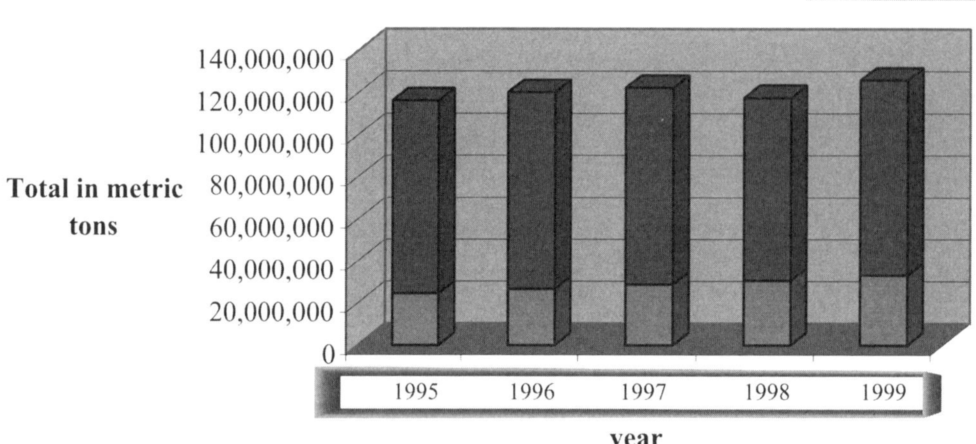

Year	World aquaculture in metric tons			World commercial catch in metric tons			Grand total
	Inland	Marine	Total	Inland	Marine	Total	
1995	14,043,180	10,449,864	24,493,044	7,264,708	84,606,586	91,871,294	116,364,338
1996	15,891,101	10,856,257	26,747,358	7,397,876	86,133,131	93,531,007	120,278,365
1997	17,485,633	11,242,532	28,728,165	7,505,685	86,260,651	93,766,336	122,494,501
1998	18,658,642	12,134,859	30,793,501	7,948,993	78,984,128	86,933,121	117,726,622
1999	20,022,908	13,287,441	33,310,349	8,260,155	84,606,398	92,866,553	126,176,902

The oceans of the Earth generate huge amounts of food for the human population. World consumption of fish and shellfish averages 34.6 pounds (15.7 kg) per person per year, and for millions of people in a variety of cultures around the world, seafood serves as the only source of protein.

World Aquaculture and Commercial Fish Catch by Species, 1998–1999

Species group	1998 Aquaculture	1998 Catch	1998 Total	1999 Aquaculture	1999 Catch	1999 Total
	---------Metric tons-------- Live-weight			---------Metric tons-------- Live-weight		
herrings, sardines, anchovies	-	16,713,557	16,713,557	-	22,715,856	22,715,856
carps, barbels, cyprinids	14,060,568	590,924	14,651,492	14,901,545	764,963	15,666,508
cods, hakes, haddocks	148	10,315,000	10,315,148	149	9,405,265	9,405,414
jacks, mullets, sauries	201,542	8,486,225	8,687,767	208,929	7,715,792	7,924,721
redfish, basses, congers	201,926	7,017,982	7,219,908	232,037	6,832,401	7,064,438
tunas, bonitos, billfishes	5,140	5,783,867	5,789,007	6,365	5,975,643	5,982,008
mackerel, snoeks, cutlassfishes	-	5,045,631	5,045,631	-	5,114,146	5,114,146
salmons, trouts, smelts	1,290,118	887,235	2,177,353	1,391,615	911,630	2,303,245
tilapias	960,370	601,075	1,561,445	1,099,268	636,164	1,735,432
flatfish	33,445	930,513	963,958	33,050	957,691	990,741
shads	-	791,021	791,021	-	842,392	842,392
sharks, rays, chimaeras	-	806,066	806,066	-	822,189	822,189
river eels	226,124	12,287	238,411	227,704	11,728	239,432
sturgeons, paddlefish	2,034	3,777	5,811	2,706	2,950	5,656
other fishes	3,068,623	15,306,125	18,374,748	3,358,417	15,922,004	19,280,421
shrimp	1,074,878	2,740,140	3,815,018	1,130,737	2,890,784	4,021,521
crabs	79,621	1,215,470	1,295,091	103,650	1,190,814	1,294,464
lobsters	40	296,024	296,064	58	312,914	312,972
krill	-	81,216	81,216	-	103,318	103,318
other crustaceans	298,741	2,044,692	2,343,433	349,497	1,788,432	2,137,929
oysters	3,539,385	160,303	3,699,688	3,711,606	157,538	3,869,144
clams	2,262,637	826,861	3,089,498	2,744,846	812,501	3,557,347
squids, cuttlefishes, octopus	33	2,633,270	2,633,303	33	3,373,463	3,373,496
mussles	1,377,631	250,134	1,627,765	1,451,032	237,823	1,688,855
scallops	874,630	522,360	1,396,990	951,866	567,507	1,519,373
abalones, winkles, conchs	3,149	95,945	99,094	2,694	105,047	107,741
other mollusks	1,123,189	2,152,395	3,275,584	1,270,001	2,093,873	3,363,874
sea urchins, other echinoderms	30	107,188	107,218	-	118,750	118,750
miscellaneous	109,499	515,838	625,337	132,544	482,975	615,519
Total	30,793,501	86,933,121	117,726,622	33,310,349	92,866,553	126,176,902

Note: Data for 1998 are revised. Data for marine mammals and aquatic plants are excluded.

Source: Personal communication from the National Marine Fisheries Service, Fisheries Statistics and Economics Division, Silver Spring, Md.

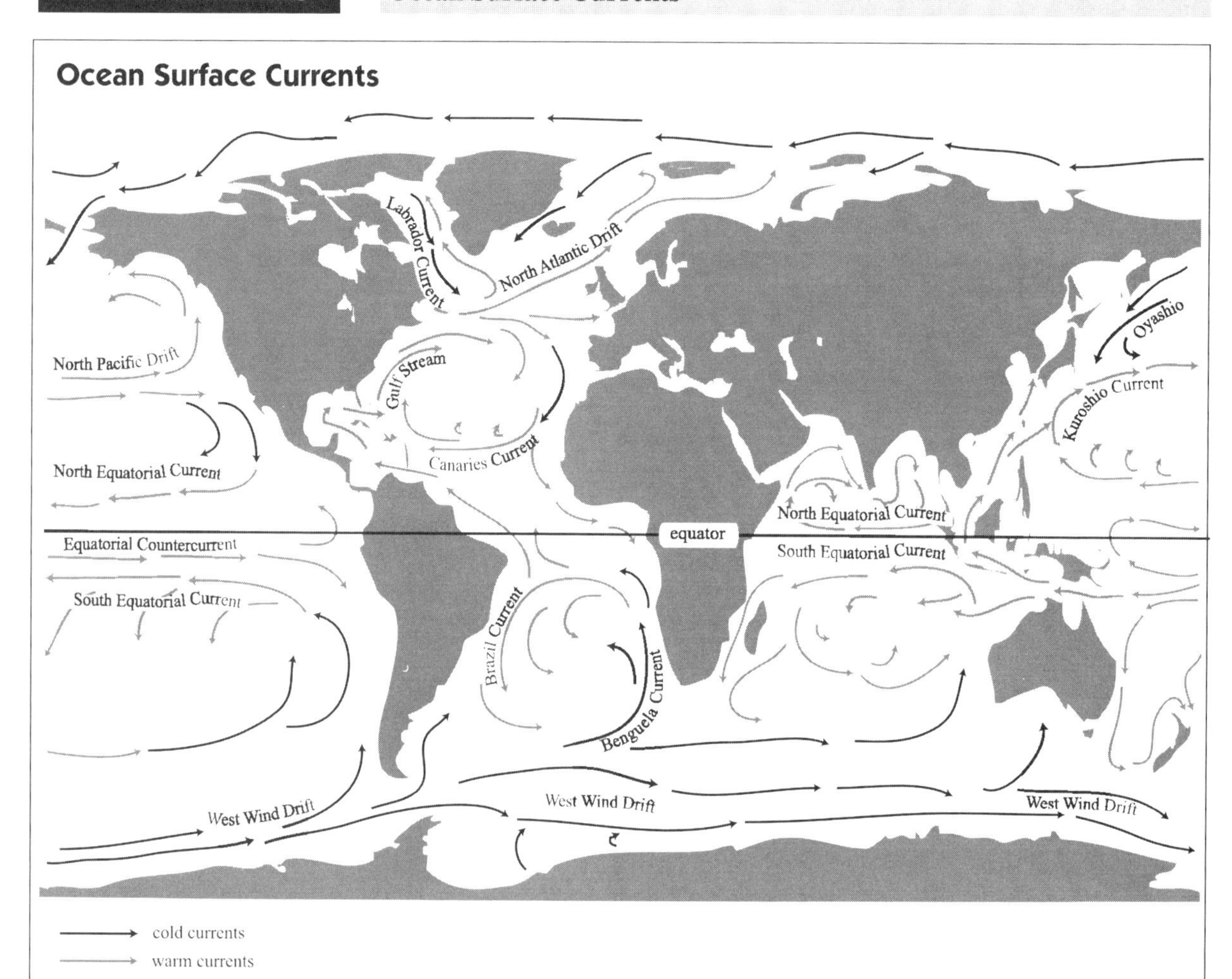

A map of world surface currents shows the general clockwise motion of the currents in the Northern Hemisphere and the general counterclockwise motion of the currents in the Southern Hemisphere, as colder water from the polar regions replaces warmer equatorial waters. This water movement carries nutrients from the polar regions to sustain marine life and to moderate global temperatures.

Global Wind Circulation

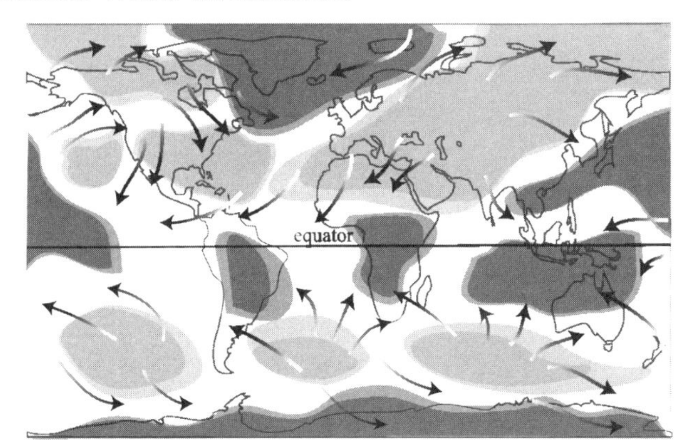

January wind pattern

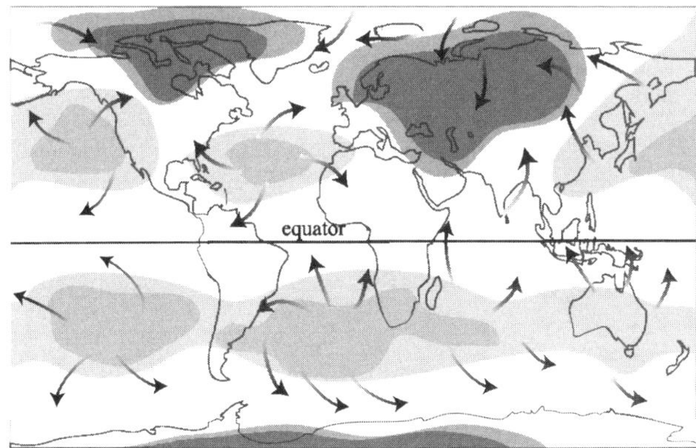

July wind pattern

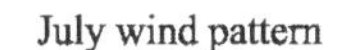

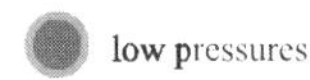

Solar radiation in the form of heat and the Earth's rotation combine to form large-scale wind patterns in the Earth's atmosphere, flowing from high to low pressure. Wind is very important to sailing vessels past and present for locomotion and is the primary generator of ocean waves.

Detailed knowledge concerning the seasonal patterns of winds is crucial to navigators when plotting courses and estimating departure and arrival times and in ensuring safe passage for all of their sailings.

Hot air rises at the equator, travels north and south as it thins out, then cools and begins to sink at the horse latitudes, where it becomes more dense, allowing it to flow back toward the equator and once again into the doldrums.

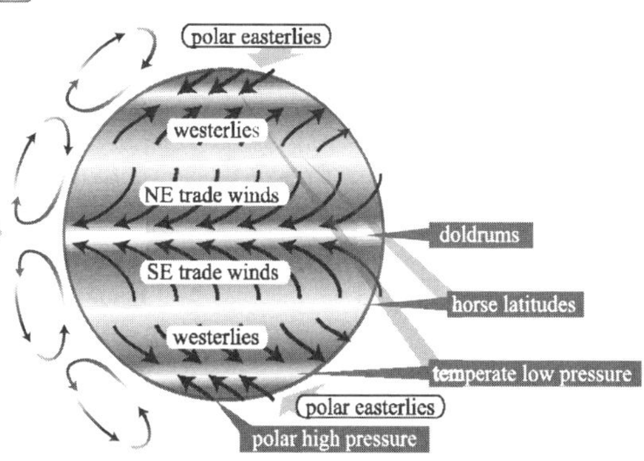

Offshore Oil Production 2001

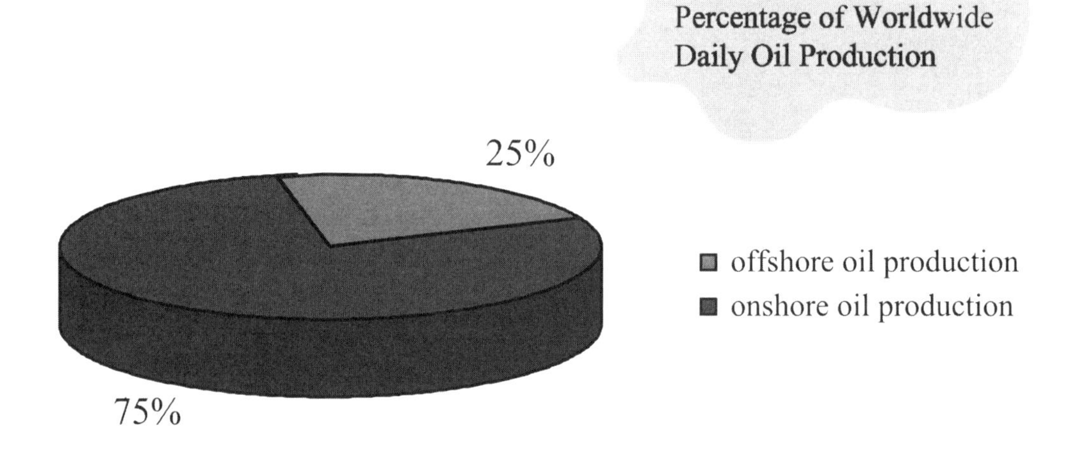

Values are in millons of barrels per day			
Year	Offshore oil production	Onshore oil production	Total oil production
2001	19	57	76

Worldwide offshore oil production is currently 25 percent of the total of world production. Offshore production of oil and gas is expected to increase substantially in the next 10 years. Other significant materials mined from the sea include phosphate minerals, large quantities of sand and gravel for construction, shell deposits containing limestone for cement production, and, recently, diamonds.

Worldwide Marine Pollution Sources

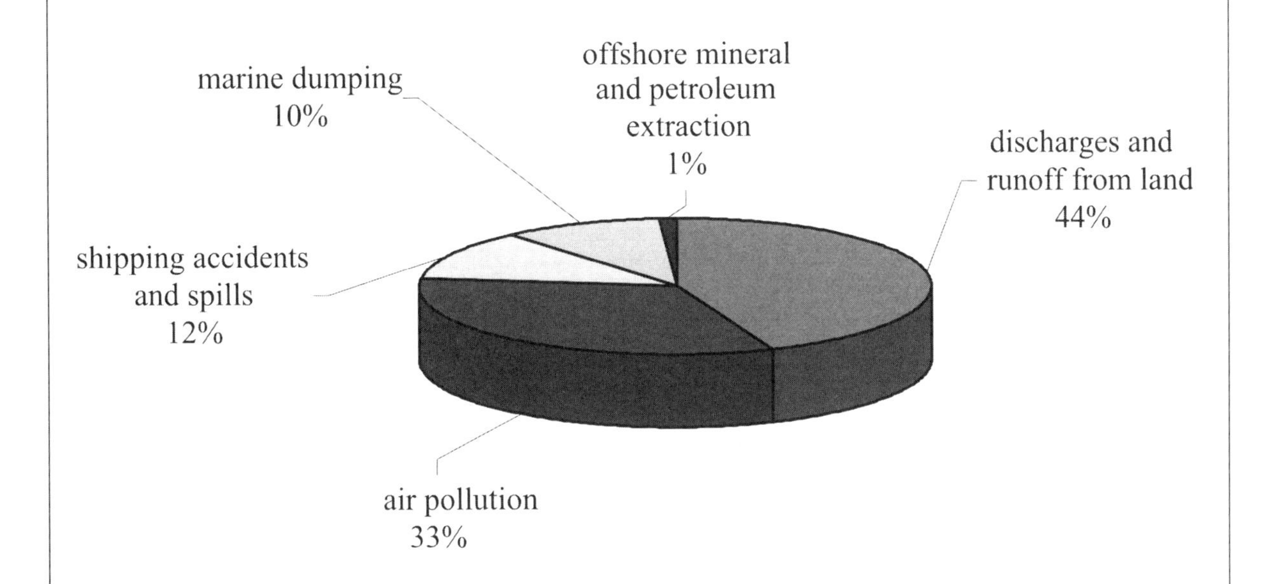

This figure illustrates that the majority of marine pollution originates in sources located on land. Although recent advances in pollution control measures and worldwide awareness have improved pollution levels, most sources continue to contaminate the marine environment. Most long-term effects of marine pollution are not known.

Relative Size of the Oceans

	Pacific Ocean	Atlantic Ocean	Indian Ocean	Southern Ocean	Arctic Ocean
Surface area (square miles)	60,068,000	29,641,000	26,473,000	7,849,000	5,428,000
Surface area (square kilometers)	155,557,000	76,762,000	68,556,000	20,327,000	14,056,000
Coastline length (miles)	84,315	69,525	41,346	11,167	28,209
Coastline length (kilometers)	135,663	111,866	66,526	17,968	45,389
Deepest point (feet)	minus 35,842	minus 28,233	minus 23,813	minus 23,738	minus 15,306
Deepest point (meters)	minus 10,924	minus 8,605	minus 7,258	minus 7,235	minus 4,665

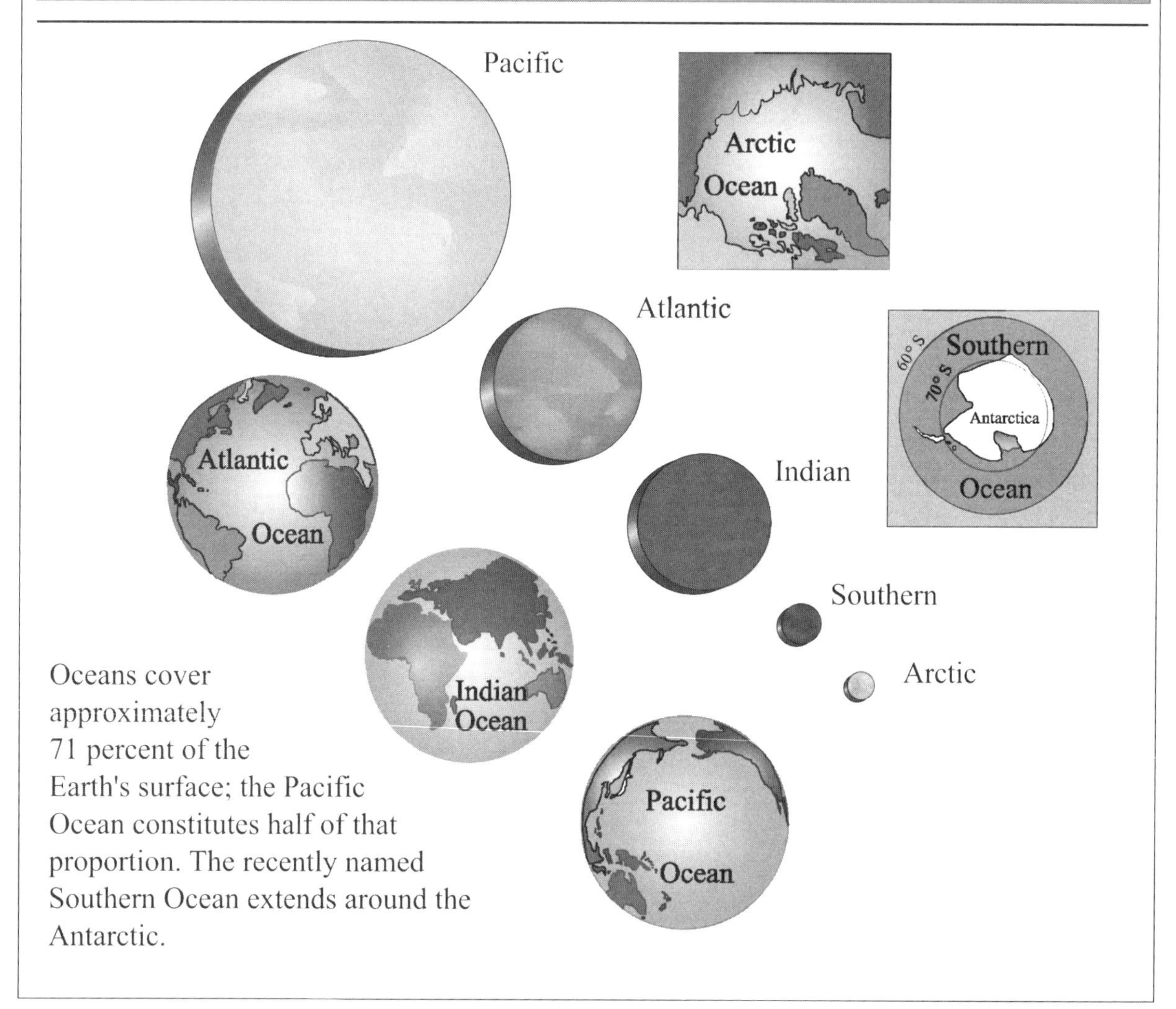

Oceans cover approximately 71 percent of the Earth's surface; the Pacific Ocean constitutes half of that proportion. The recently named Southern Ocean extends around the Antarctic.

CHARTS & TABLES Relative Size of the Oceans

Major World Shipping Routes

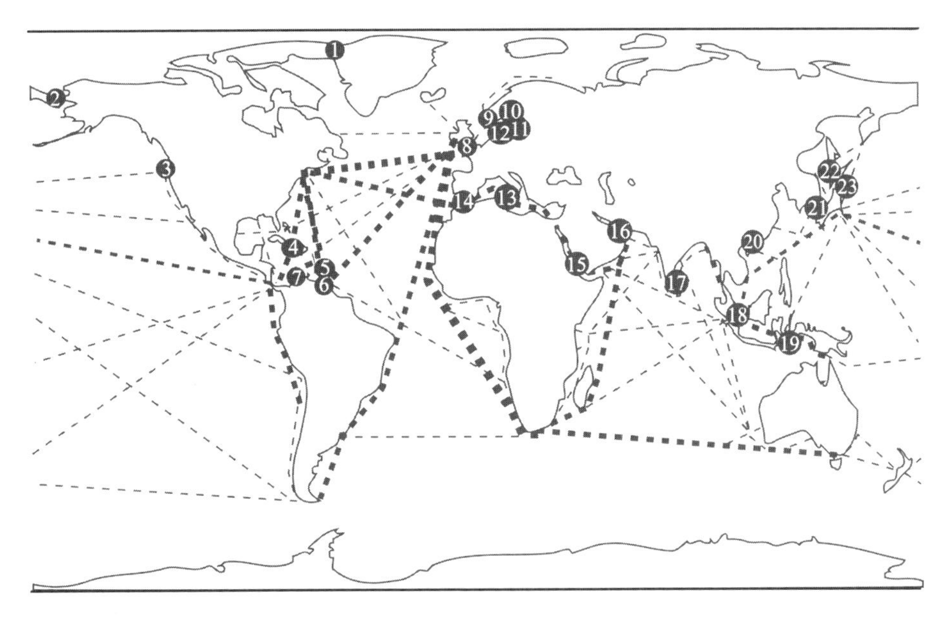

- - - - - - - - less heavily traveled

▪▪▪▪▪▪▪▪ more heavily traveled

The oceans have provided major transportation routes for worldwide commerce since civilization began. Over 28,000 merchant ships sail these shipping routes annually, carrying a vast array of commercial items: oil, gas, cars, trucks, agricultural products, mining ores, and metals. Daily, these ships pass through major international straits. It is a matter of great importance to vessels whether the waterway is free or under the control of the countries that flank it; if the latter, transit fees must be paid.

International Straits

1. Robeson Channel
2. Bering Strait
3. Juan de Fuca Strait
4. Windward Passage
5. Martinique Passage
6. Saint Lucia Channel
7. Aruba-Paraguana Passage
8. Strait of Dover
9. Øresund
10. Into Gulf of Bosnia
11. Into Gulf of Finland
12. Bornholmsgattet
13. Bonifacio Strait
14. Strait of Gibraltar
15. Bab al Mandab
16. Strait of Hormuz
17. Palk Strait
18. Malacca Strait
19. Strait of Ombai, Selat
20. Lema Channel
21. Chosen Strait
22. Soya Strait
23. Notsuki-Suido

Tides

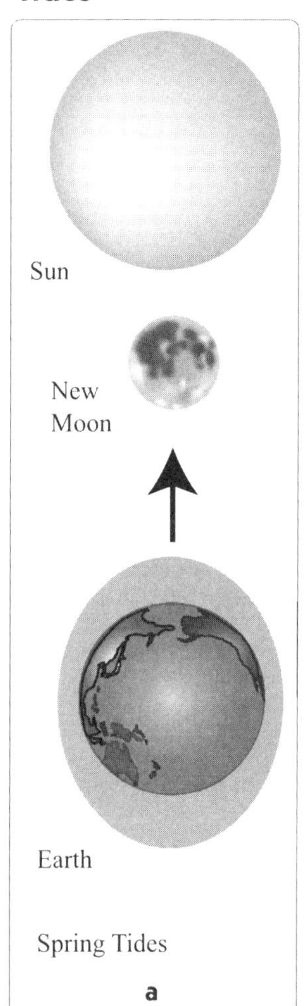

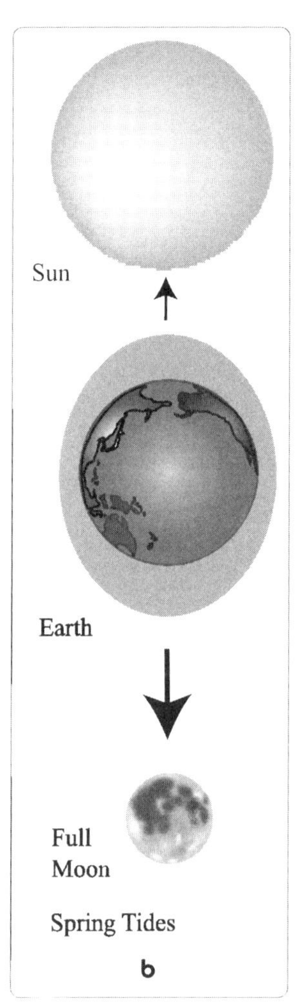

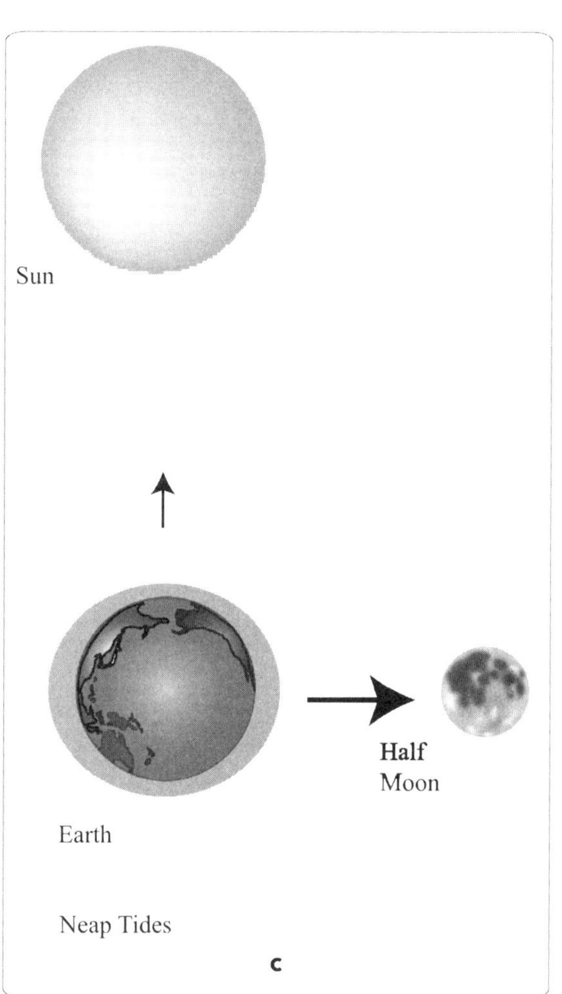

Ocean tides are caused by gravitational forces exerted by the Sun and Moon on the Earth. Large tidal changes are visible over just a few hours along many miles of coastline throughout the world; some daily tidal ranges exceed 40 feet (12.2 m). These particularly large tides, called spring tides, occur when the Sun and Moon are in alignment, as in **a** and **b**. Neap tides, in **c**, are tides with less daily tidal range that occur at times when the Moon and Sun are at right angles to each other or, in other words, at quadrature.

Sample Tide Chart

July 2003, Juneau, Alaska (Pacific)

Date	Time	Ht.	Time	Ht.	Time	Ht.	Time	Ht.
7/1/2003	2:29 A.M. H	16.7	9:07 A.M. L	-2.2	3:38 P.M. H	14.6	9:09 P.M. L	3.7
7/2/2003	3:07 A.M. H	16.6	9:44 A.M. L	-2.1	4:15 P.M. H	14.8	9:50 P.M. L	3.7
7/3/2003	3:47 A.M. H	16.2	10:23 A.M. L	-1.7	4:54 P.M. H	14.9	10:35 P.M. L	3.7
7/4/2003	4:31 A.M. H	15.6	11:04 A.M. L	-1.0	5:35 P.M. H	15.0	11:25 P.M. L	3.6
7/5/2003	5:21 A.M. H	14.7	11:48 A.M. L	0.0	6:20 P.M. H	15.1		
7/6/2003	12:24 A.M. L	3.4	6:20 A.M. H	13.7	12:39 P.M. L	1.1	7:10 P.M. H	15.4
7/7/2003	1:31 A.M. L	2.9	7:30 A.M. H	12.8	1:36 P.M. L	2.2	8:05 P.M. H	15.7
7/8/2003	2:43 A.M. L	2.0	8:49 A.M. H	12.3	2:41 P.M. L	3.2	9:04 P.M. H	16.1
7/9/2003	3:53 A.M. L	0.8	10:11 A.M. H	12.5	3:50 P.M. L	3.7	10:04 P.M. H	16.7
7/10/2003	4:58 A.M. L	-0.7	11:24 A.M. H	13.2	4:55 P.M. L	3.7	11:04 P.M. H	17.3
7/11/2003	5:55 A.M. L	-2.0	12:27 P.M. H	14.1	5:55 P.M. L	3.4		
7/12/2003	12:01 A.M. H	18.0	6:47 A.M. L	-3.1	1:21 P.M. H	15.0	6:49 P.M. L	2.9
7/13/2003	12:54 A.M. H	18.4	7:35 A.M. L	-3.8	2:10 P.M. H	15.7	7:39 P.M. L	2.5
7/14/2003	1:44 A.M. H	18.6	8:21 A.M. L	-4.0	2:55 P.M. H	16.1	8:27 P.M. L	2.2
7/15/2003	2:32 A.M. H	18.4	9:04 A.M. L	-3.6	3:38 P.M. H	16.2	9:13 P.M. L	2.1
7/16/2003	3:17 A.M. H	17.8	9:45 A.M. L	-2.9	4:19 P.M. H	16.1	9:58 P.M. L	2.3
7/17/2003	4:01 A.M. H	17.8	10:25 A.M. L	-1.8	5:00 P.M. H	15.8	10:44 P.M. L	2.7
7/18/2003	4:45 A.M. H	15.6	11:05 A.M. L	-0.5	5:40 P.M. H	15.3	11:32 P.M. L	3.1
7/19/2003	5:31 A.M. H	15.6	11:45 A.M. H	1.1	6:21 P.M. H	14.8		
7/20/2003	12:24 A.M. L	3.4	6:22 A.M. H	2.8	12:28 P.M. L	2.6	7:05 P.M. H	14.3
7/21/2003	1:23 A.M. L	3.7	7:22 A.M. H	11.6	1:17 P.M. L	4.0	7:52 P.M. H	13.9
7/22/2003	2:29 A.M. L	3.6	8:34 A.M. H	10.8	2:16 P.M. L	5.1	8:46 P.M. H	13.7
7/23/2003	3:36 A.M. L	3.1	9:55 A.M. H	10.7	3:23 P.M. L	5.8	9:44 P.M. H	13.7
7/24/2003	4:38 A.M. L	2.2	11:09 A.M. H	11.2	4:29 P.M. L	5.9	10:41 P.M. H	14.2
7/25/2003	**5:31 A.M. L**	**1.2**	**12:07 P.M. H**	**11.9**	**5:26 P.M. L**	**5.5**	**11:32 P.M. H**	**14.8**
7/26/2003	6:16 A.M. L	0.1	12:52 P.M. H	12.8	6:14 P.M. L	4.9		
7/27/2003	12:18 A.M. H	15.6	6:57 A.M. L	-0.9	1:31 P.M. H	13.7	6:56 P.M. L	4.2
7/28/2003	12:59 A.M. H	16.3	7:35 A.M. L	-1.8	2:06 P.M. L	4.5	7:36 P.M. L	3.5
7/29/2003	1:38 A.M. H	16.9	8:11 A.M. L	-2.4	2:40 P.M. H	15.2	8:15 P.M. L	2.8
7/30/2003	2:16 A.M. H	17.3	8:47 A.M. L	-2.8	3:13 P.M. H	15.7	8:54 P.M. L	2.3
7/31/2003	2:55 A.M. H	17.4	9:23 A.M. L	-2.7	3:47 P.M. H	16.2	9:34 P.M. L	1.8

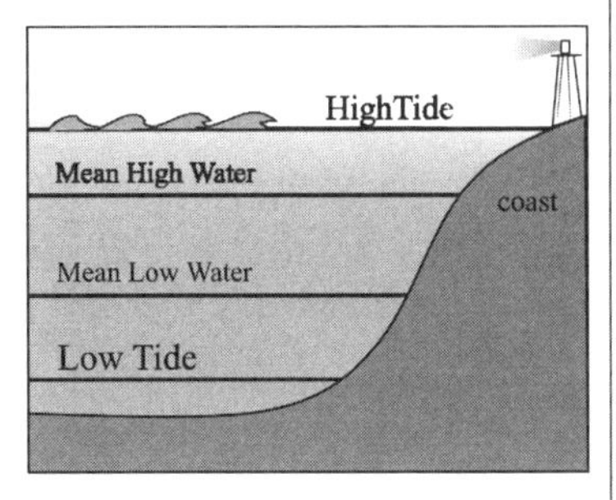

Atlantic high and low tide measurement

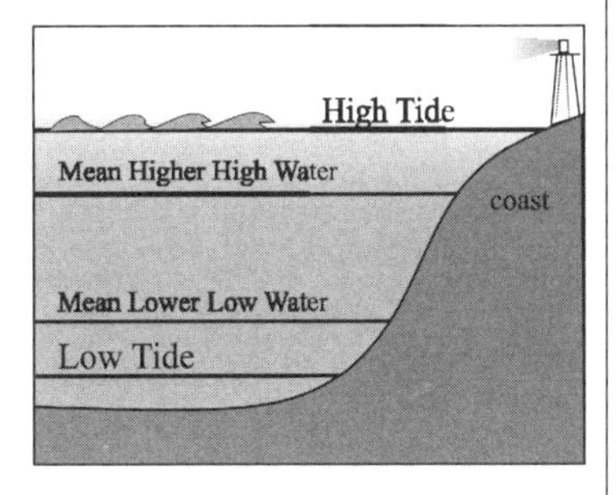

Pacific high and low tide measurement

Multiply feet by 0.3048 to obtain meters

L = Low Tide H = High Tide

Tide tables can usually be obtained for most coastal regions of the Earth to supply mariners with vital depth information. The most common tide has two high waters and two low waters per day. High tides are measured from the average of the high tides, or mean high water (mhw), and low tides are measured from the average of the low tides, or mean low water (mlw).

Generally in the Pacific Ocean, there are large differences in height between the two high and low tides; thus the high tide is measured from the mean higher high water (mhhw) and the low tide is measured from the mean of the lower low water (mllw). On the **boldfaced** date, July 25, 2003, an example for Juneau, Alaska, shows that a low tide of 1.2 feet above mean lower low water occurs at 5:31 A.M. local time. At 12:07 P.M., 6 hours 36 minutes later, a high tide of 11.9 feet above mean higher high water occurs. Another low tide (mllw) of 5.5 feet occurs at 5:26 P.M., and a high tide (mhhw) of 14.8 feet follows at 11:32 P.M.

Examples of Two Ocean Food Chains

Pelagic food chain

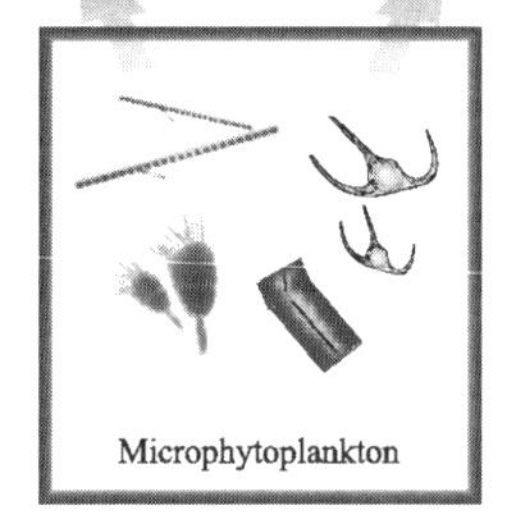

Neritic food chain

Ocean food chains contain huge quantities of living organisms, such as the Pacific plankton, which has been in existence for around 26 million years. Energy essential to life is transferred organism by organism, from the planktons on up through the food chain to the larger fishes.

Food Web of Southern Oceans

In some food chains, a food web is a more accurate description of the relationships among species because prey can sometimes eat different species of food at different levels of the chain, such as in the food web of the southern latitudes.

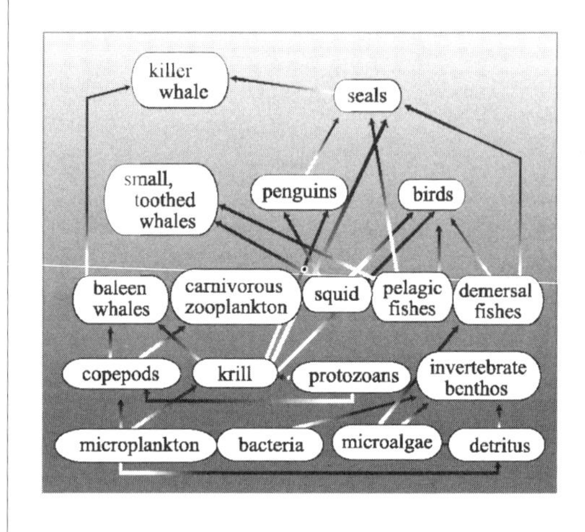

Beaufort Wind Scale

Beaufort code value	Beaufort's scale	Appearance of sea surface	Wind velocity in mph	Wind velocity in kph	Wind velocity in knots	Weather forecast Term
0.	CALM	Sea appears mirrorlike.	less than 1	less than 1	less than 1	Calm
1.	LIGHT AIR	Small ripples with no white foam crests.	1–3	2–6	1–3	Light
2.	LIGHT BREEZE	Wavelets appear; crests do not break.	4–7	7–12	4–6	Light
3.	GENTLE BREEZE	Larger wavelets with occasional whitecaps appear.	8–12	13–19	7–10	Gentle
4.	MODERATE BREEZE	Smaller waves become longer with frequent whitecaps.	13–18	20–30	11–16	Moderate
5.	FRESH BREEZE	Larger waves form with whitecaps and spray.	19–24	31–39	17–21	Fresh
6.	STRONG BREEZE	Whitecaps are everywhere; there is some spray.	25–31	40–50	22–27	Strong
7.	MODERATE GALE	Streaking white foam appears on the back of breaking waves.	32–38	51–61	28–33	Strong
8.	FRESH GALE	Larger waves become higher with spindrift on breaking crests and extensive foam.	39–46	62–75	34–40	Strong
9.	STRONG GALE	Sea begins to roll with high waves and heavy foam.	47–54	76–88	41–47	Gale
10.	WHOLE GALE	Very high waves occur and the sea appears white.	55–63	89–102	48–55	Gale
11.	STORM	Exceptionally high waves occur. Small and medium-sized ships are not visible between wave crests.	64–72	103–118	56–63	Whole gale
12.	HURRICANE	The sea is completely white with driving spray. Visibility is poor and the air is filled with foam and spray.	73–82	119–132	64–71	Hurricane
13.	HURRICANE		83–92	133–148	72–80	Hurricane
14.	HURRICANE		93–102	149–165	81–89	Hurricane
15.	HURRICANE		103–114	166–184	90–99	Hurricane
16.	HURRICANE		115–125	185–202	100–109	Hurricane
17.	HURRICANE		126–136	203–219	110–118	Hurricane

The Beaufort wind scale is a scale of wind forces developed in1805 by the British admiral Sir Francis Beaufort. Devised to describe the appearance of the sea at successively higher wind speeds, the scale is still the most convenient method of classifying wind and sea conditions.

Prefixes and Suffixes Relating to the Marine Environment Derived from Latin (L) and Greek (G)

abysso- deep (G)
aero- air, atmosphere (G)
alga- seaweed (L)
ampho- both, double (G)
anomal- irregular,uneven (G)
aqua- water (L)
arch- first in time, beginning (G)
arena- sand (L)
arthro- joint (G)
astro- star (G)
bas- base, bottom (G)
batho- deep (G)
bio- living (G)
branchi- gilllike (G)
bryo- moss (G)
calic- cup (L)
calori- heat (L)
capill- hair (L)
capit- head (L)
caud- tail (L)
cephalo- head (G)
cer-, cera- horn (G)
chiton- tunic (G)
chlor- green (G)
chondri- cartilage (G)
chrom- color (G)
cirri- hair (L)
clino- slope (G)
cornu- horn (L)
cran- skull (G)
crusta- shell (L)
-cyst bladder, bag (G)
-cyte cell (G)
dactyl- finger (G)
deca- ten (L)
dent- tooth (L)
derm- skin (G)
di- two, double (G)
dia- through, across (G)
dur- hard (L)
echino- spiny (G)
eco- house, abode (G)
ecto- outside, outer (G)
endo- inner, within (G)
exo- out, without (G)
fecund- fruitful (L)
fluvi- river (L)
geno- birth, race (G)
geo- earth (G)
giga- very large (G)
gyr- round, circle (G)
halo- salt (G)
helio-sun (G)
hemi- half (G)
herbi- plant (L)
holo- whole (G)
hydro- water (G)
hygro- wet (G)
icthyo- fish (G)
-idae members of the animal family (L)
infra- below, beneath (L)
insula- island (L)
inter- between (L)
intr- inside (L)
iso- equal (G)
-ite signifying a mineral or rock (G)
lamino- layer (L)
lati- broad, wide (L)
litho- stone (G)
lorica- armor (L)
luna- moon (L)
macro- large (G)
magn- great, large (L)
mari- sea (L)
medi- middle (L)

(continues)

Prefixes and Suffixes Relating to the Marine Environment Derived from Latin (L) and Greek (G) *(continued)*

medus- jellyfish (G)
mega- great, large (G)
meso- middle (G)
meta- after (G)
-meter- measure (G)
-metry science of measuring (G)
micro- small (G)
mono- one, single (G)
-morph form (G)
nano- dwarf (G)
necto- swimming (G)
neo- new (G)
-nomy the science of (G)
ob- opposite (L)
octo- eight (L/G)
oculo- eye (L)
odonto- teeth (G)
-ology science of (G)
omni- all (L)
opto- eye, vision (G)
-osis signifying a process (G)
oste- bone (G)
ovo- egg (L)
paleo- ancient (G)
pari- equal (L)
pedi- foot (L)
penta- five (G)
photo- light (G)
phyto- plant (G)
pisci- fish (L)
platy- broad, wide (G)
plankto- wandering (G)
pluri- several (L)
pneuma- air, breath (G)
pod- foot (G)
poly- many (G)
pre- before (L)
primo- first (L)
proto- first (G)
pulmo- lung (L)
quarda- four (L)
sali- salt (L)
sarc- flesh (G)
scler- hard (G)
semi- half (L)
strati- layer (L)
sub- below (L)
supra- above (L)
tele- from afar (G)
terra- earth (L)
thalasso- sea (G)
theri- wild animal (G)
therm- heat (G)
tri- three (G)
tunic- cloak or covering (L)
ultim- farthest, last (L)
vas- vessel (L)
-vorous feeding on (L)
zoo- animal (G)
zyg- pair (G)

Recommended Reading and Useful Websites

Anikouchine, William A., Richard W. Sternberg, et al. *The World Ocean,* 2nd ed. Englewood Cliffs, N.J.: Prentice-Hall, 1981.

Baker, Lucy. *Life in the Oceans.* New York: Scholastic, 1993.

Belleville, Bill. *Deep Cuba: The Inside Story of an American Oceanographic Expedition.* Athens: University of Georgia Press, 2002.

Bird, Jonathan. *Beneath the North Atlantic.* Hartford, Conn.: Tidemark 1997.

Brandon, Jeffrey L., and Frank Rokop. *Life between the Tides.* San Diego, Calif.: American Southwest Publishing, 1985.

Cassie, Brian., et al. *National Audubon Society First Field Guide: Shells.* New York: Scholastic and Chanticleer Press, 2000.

Charton, Barbara. *The Facts On File Dictionary of Marine Science,* rev. ed. New York: Facts On File, 2001.

Colin, Patrick L. *Marine Invertebrates and Plants of the Living Reef.* Neptune, N.J.: TFH Publications, 1990.

Cousteau, Jacques. *Jacques Cousteau: The Ocean World.* New York: Harry N. Abrams, 1992.

Deacon, G. E. R., et al. *Seas, Maps and Men.* New York: Doubleday, 1962.

The Diagram Group. *The Facts On File Earth Science Handbook.* New York: Facts On File, 2000.

Earle, Sylvia A. *The National Geographic Atlas of the Ocean: The Deep Frontier.* Washington, D.C.: National Geographic Society, 2001.

Ellis, Richard. *Deep Atlantic: Life, Death, and Exploration in the Abyss.* New York: Knopf, 1996.

———. *Encyclopedia of the Sea.* New York: Knopf, 2000.

Erickson, Jon. *Marine Geology: Exploring the New Frontiers of the Ocean,* rev. ed. New York: Facts On File, 2002.

Folkens, Pieter, illus. *National Audubon Society's Guide to Marine Mammals of the World.* New York: Knopf, 2002.

Hine, Robert, et al. *The Facts On File Dictionary of Biology,* 3rd ed. New York: Facts On File, 1999.

Horn, Michael H., Karen L. M. Martin, and Michael A. Chotkowski, eds. *Intertidal Fishes: Life in Two Worlds.* San Diego, Calif.: Academic Press, 1998.

Kennish, Michael J. *Practical Handbook of Marine Science,* 2nd ed. Boca Raton, Fla.: CRC Press, Inc., 1994.

Klimely, A. Peter, ed., and David G. Ainley. *Great White Sharks: The Biology of Carcharodon Carcharias.* San Diego, Calif.: Academic Press, 1998.

Lambert, David. *The Oceans.* New York: Warwick Press, 1979.

Mauseth, James D., et al. *Botany,* 2nd ed. Philadelphia: Saunders College Publishing, 1995.

Microsoft Encarta Encyclopedia. Available online. URL: www.encarta.msn.com
Milne, David H., and A.A. Milne. *Marine Life and the Sea.* Belmont, Calif.: Wadsworth Publishing, 1995.
The Random House Atlas of the Oceans. New York: Random House, 1991.
Ricketts, Edward Flanders, Joel W. Hedgepeth, and Jack Calvin. *Between Pacific Tides,* 5th ed. Stanford, Calif.: Stanford University Press, 1986.
Thorne-Miller, Boyce, and Sylvia A. Earle. *The Living Ocean: Understanding and Protecting Marine Biodiversity,* 2nd ed. Washington, D. C.: Island Press, 1998.
Tomas, Carmelo R., Grethe R. Hasle, Jahn Throndsen, and Beit Heimdal, eds. *Identifying Marine Phytoplankton.* San Mateo, Calif.: Morgan Kaufmann Publishers, 1997.
Van Dorn, William G. *Oceanography and Seamanship.* New York: Dodd, Mead, 1974.
Ville, Claude A., Warren F. Walker, Jr., Robert D. Barnes, et al. *General Zoology,* 4th ed. Philadelphia: W. B. Saunders, 1973.
Waller, Geoffrey. *Sea Life.* Washington, D.C.: Smithsonian Institution Press, 1996.
World Book Encyclopedia. Chicago: World Book, 1998.

Useful Websites

Alaska Fisheries and Science Center: www.afsc.noaa.gov
Alfred Wegener Institute: http://www.awi-bremerhaven.de/#
Alliance for Marine Remote Sensing: www.waterobserver.org
American Society of Limnology and Oceanography: http://aslo.org
Argentine Oceanographic Data Center: www.conae.gov.ar/~ceado
Atlantic Oceanographic and Meteorological Laboratory: www.aoml.noaa.gov
Australian Oceanographic Data Center: www.aodc.gov.au/AODC.html
Bigelow Laboratory of Ocean Sciences: http://www.bigelow.org
British Oceanographic Data Center: http://www.bodc.ac.uk
Center for Sea and Atmosphere Research: www-cima.at.fcen.uba.ar
Coastal Ocean Probing Experiment: www.etl.noaa.gov/review/os/kropfli
Coastal Zone Management Program: http://www.ocrm.nos.noaa.gov/czm
CoastWatch: http://coastwatch.noaa.gov
Consortium for Oceanographic Research and Education: http://core.cast.msstate.edu
Coordinated Eastern Arctic Experiment: www-nsidc.colorado.edu/data/nsidc-0020.html
FishBase A worldwide database for information on fish: http://www.fishbase.org

General Bathymetric Chart of the Oceans: www.ngdc.noaa.gov/mgg/gebco/gebco.html
Global Drifter Center: www.aoml.noaa.gov/phod/dac/gdc.html
Global Hydrology and Climate Center: www.ghcc.msfc.nasa.gov
Hakluyt Society Books on maritime and geographical discovery: www.hakluyt.com
Intergovernmental Oceanographic Commission: ioc.unesco.org/iocweb
International Association for the Physical Sciences of the Ocean: http://www.olympus.net/IAPSO
International Council for the Exploration of the Sea: http://www.ices.dk
International Council of Scientific Unions: www.icsu.org
International Hydrographic Organization: www.iho.shom.fr
International Oceanographic Data and Information Exchange: ioc.unesco.org/iode
International Whaling Commission: www.marine.gov.uk/iwc.htm
Joint Institute for the Study of Atmosphere and the Ocean: www.jisao.washington.edu
Marine Technology Society: www.mtsociety.org
National Academy of Sciences: www.nationalacademies.org/nas
National Sea Grant Library: http://nsgl.gso.uri.edu/diglib.html.
National Oceanic and Atmospheric Administration: www.noaa.gov
The Oceanography Society: www.tos.org
Pelagic Fisheries Research Program: www.soest.hawaii.edu/PFRP/pfrp1.html
ReefBase, A global information system on coral reefs: www.reefbase.org
Scripps Institution of Oceanography: http://www.sio.ucsd.edu
Seaworld (Division of SAAMBR): www.seaworld.org.za
TOPEX/Poseidon Cooperative project between the U.S and France to observe the earth's oceans: topex-www.jpl.nasa.gov
United Nations Educational, Scientific and Cultural Organization: www.unesco.org
University-National Oceanographic Laboratory Systems: www.unols.org
Wildlife Conservation Society: www.wcs.org
Woods Hole Oceanographic Institution: http://www.whoi.edu

Italic page numbers indicate illustrations or captions.